Also available in the Inverse and Ill-Posed Problems Series:

Linear Sobolev Type Equations and Degenerate Semigroups of Operators
G.A. Sviridyuk and V.E. Fedorov
Ill-Posed and Non-Classical Problems of Mathematical Physics and Analysis
Editors: M.M. Lavrent'ev and S.I. Kabanikhin
Forward and Inverse Problems for Hyperbolic, Elliptic and Mixed Type Equations
A.G. Megrabov
Nonclassical Linear Volterra Equations of the First Kind
A.S. Apartsyn
Poorly Visible Media in X-ray Tomography
D.S. Anikonov, V.G. Nazarov, and I.V. Prokhorov
Dynamical Inverse Problems of Distributed Systems
V.I. Maksimov
Theory of Linear Ill-Posed Problems and its Applications
V.K. Ivanov, V.V. Vasin and V.P. Tanana
Ill-Posed Internal Boundary Value Problems for the Biharmonic Equation
M.A. Atakhodzhaev
Investigation Methods for Inverse Problems
V.G. Romanov
Operator Theory. Nonclassical Problems
S.G. Pyatkov
Inverse Problems for Partial Differential Equations
Yu.Ya. Belov
Method of Spectral Mappings in the Inverse Problem Theory
V. Yurko
Theory of Linear Optimization
I.I. Eremin
Integral Geometry and Inverse Problems for Kinetic Equations
A.Kh. Amirov
Computer Modelling in Tomography and Ill-Posed Problems
M.M. Lavrent'ev, S.M. Zerkal and O.E. Trofimov
An Introduction to Identification Problems via Functional Analysis
A. Lorenzi
Coefficient Inverse Problems for Parabolic Type Equations and Their Application
P.G. Danilaev
Inverse Problems for Kinetic and Other Evolution Equations
Yu.E. Anikonov
Inverse Problems of Wave Processes
A.S. Blagoveshchenskii
Uniqueness Problems for Degenerating Equations and Nonclassical Problems
S.P. Shishatskii, A. Asanov and E.R. Atamanov
Uniqueness Questions in Reconstruction of Multidimensional Tomography-Type Projection Data
V.P. Golubyatnikov
Monte Carlo Method for Solving Inverse Problems of Radiation Transfer
V.S. Antyufeev
Introduction to the Theory of Inverse Problems
A.L. Bukhgeim
Identification Problems of Wave Phenomena - Theory and Numerics
S.I. Kabanikhin and A. Lorenzi
Inverse Problems of Electromagnetic Geophysical Fields
P.S. Martyshko
Composite Type Equations and Inverse Problems
A.I. Kozhanov
Inverse Problems of Vibrational Spectroscopy
A.G. Yagola, I.V. Kochikov, G.M. Kuramshina and Yu.A. Pentin
Elements of the Theory of Inverse Problems
A.M. Denisov
Volterra Equations and Inverse Problems
A.L. Bughgeim
Small Parameter Method in Multidimensional Inverse Problems
A.S. Barashkov
Regularization, Uniqueness and Existence of Volterra Equations of the First Kind
A. Asanov
Methods for Solution of Nonlinear Operator Equations
V.P. Tanana
Inverse and Ill-Posed Sources Problems
Yu.E. Anikonov, B.A. Bubnov and G.N. Erokhin
Methods for Solving Operator Equations
V.P. Tanana
Nonclassical and Inverse Problems for Pseudoparabolic Equations
A. Asanov and E.R. Atamanov
Formulas in Inverse and Ill-Posed Problems
Yu.E. Anikonov
Inverse Logarithmic Potential Problem
V.G. Cherednichenko
Multidimensional Inverse and Ill-Posed Problems for Differential Equations
Yu.E. Anikonov
Ill-Posed Problems with A Priori Information
V.V. Vasin and A.L. Ageev
Integral Geometry of Tensor Fields
V.A. Sharafutdinov
Inverse Problems for Maxwell's Equations
V.G. Romanov and S.I. Kabanikhin

Inverse and Ill-Posed Problems Series

Ill-Posed Boundary-Value Problems

S.E. Temirbolat

///VSP///
Utrecht • Boston
2003

VSP
an imprint of Brill Academic Publishers
P.O. Box 346
3700 AH Zeist
The Netherlands

Tel: +31 30 692 5790
Fax: +31 30 693 2081
vsppub@compuserve.com
www.brill.nl
www.vsppub.com

First published in 2003

ISBN 90-6764-395-5

Printed in The Netherlands by Ridderprint bv, Ridderkerk.

Preface

The term "ill-posed problems" used to mean problems with unstable solutions, for which small errors in initial data lead to a significant error in the results. These problems had been considered unphysical for a long time and therefore of no serious interest for researchers. It was a common viewpoint even after the publications by Novikov (1938) and Tikhonov (1943), which emphasized the practical significance of such problems and showed the possibility of their stable solution. Starting from the late 1950s, new practically important problems have been formulated, which expanded the notion of ill-posed problems, and new approaches to their study have emerged.

At present there are numerous articles, books and other publications devoted to the theory of ill-posed problems, and the theory itself has been sufficiently well-developed. At the same time the topic is by no means exhausted. This study is an attempt at extending well-known facts to new classes of problems and at working out novel approaches to the solution of these problems. To be more exact, the monograph is devoted to the questions of [consistency] of ill-posed boundary-value problems for systems of various types of the first-order differential equations with constant coefficients and to the methods for their solution.

The book mostly contains the results obtained by the author and the findings of other authors are shown only in connection with the material being presented.

The introductory chapter gives some facts of the theory of matrices and ordinary differential equations necessary for understanding of the book. In particular, direct and inverse problems of canonical forms are studied and the relation between an arbitrary ordinary equation and a first-order system.

The first chapter studies the well posedness of the problem for an arbitrary system of ordinary equations and the occurrence of ill-posed problems among them.

In the second chapter the similar approach is applied to various kinds of parabolic systems and the solution methods for ill-posed problems are given.

In the third chapter the above-described methods are used to study problems for hyperbolic-type equations.

Some parts of the book were presented as special courses and seminars for undergraduate, graduate and post-graduate students of the department of mechanics and mathematics at the Al-Farabi National Kazakh University.

The book is intended for a reader basically informed about differential equations and theory of boundary-value problems.

The author is well aware of the complexity of the task he undertook and is grateful in advance for all justified critical remarks.

Contents

Introduction

1. CONDITIONALLY WELL-POSED PROBLEMS AND RELATED QUESTIONS

Every problem of mathematical physics can be written in the operator form as

$$Ax = y, \qquad x \in D_A \subseteq X, \quad y \in Y, \tag{1.1}$$

where the operator A with a non-empty domain of definition D_A acts from the function space X onto the similar space Y.

Pre-specification of the spaces X and Y is the necessary element of the mathematical statement of problem (1.1)) and its well-posedness is determined in close connection with its well-posedness. Two notions of well-posedness are known in special literature for the problem statement of mathematical physics. The first of them was introduced by the eminent mathematician J. Hadamard (1902, 1952).

Problem (1.1) is called well-posed if the following three conditions are satisfied:

- the range space R_A of the operator A coincides with the space Y (solvability);
- the equality $Ax_1 = Ax_2$ for x_1 and x_2 from D_A implies the equality $x_1 = x_2$ (uniqueness);
- the inverse operator A^{-1} exists and is continuous on Y (stability).

The fulfillment of these conditions seemed so natural for any problem within reason that Hadamard believed the problem, for which these conditions are violated, to be non-physical.

Naturally, such problems were considered [inadequately stated or simply] ill-posed. In other words, a problem is traditionally considered an ill-posed one either if it has a non-unique solution, or if it is solvable not for every type of initial data, or if it has a non-unique solution, or if it is impossible to match the norms of solution and norms of initial data to secure continuous dependence between those.

As an example of an ill-posed problem Hadamard mentions the Cauchy problem for the Laplace equation with a violated stability condition. It was later proved that many problems of theoretical and practical significance are reduced to this problem. In particular, the matters of extension of analytical and harmonic functions as well as some geophysical and biophysical problems are tantamount to solving the Cauchy problem for the Laplace equation. Therefore, this problem has attracted considerable attention of researchers (Lavrent'ev, 1956; John, 1955) since the early 1950s.

Hadamard proclaimed nonphysicality of the ill-posed problem, which, we believe, had hindered this direction of research for some time. The first serious result in this field was obtained as late as in 1938. It was the theorem about the solution uniqueness of the inverse problem of the potential theory proved by Novikov (1938) at the request of geophysicists. This was followed by the work of Tikhonov (1943) introducing an important class of inverse problems associated with the reconstruction of the characteristics of the studied medium from the measurable physical fields generated by it and proving the stability of their solution.

The second, generalizing or classical (or in Hadamard's sense) notion of well-posedness based on A. N. Tikhonov's fundamental idea of the restricting of the domain of definition of the operator A, was formulated by Lavrent'ev (1962).

Problem (1.1) is called well-posed in Tikhonov's sense or conditionally well-posed provided:

— it is known a priori that the solution of problem (1.1) exists for a certain class N of data from the space Y and it belongs to a certain set $M \subset D_A$;

— the solution is unique on the set M;

— to small changes in the right-hand side y not causing the solution to go outside the set M there correspond small solution variations (conditional stability).

The set M is called the set of well-posedness.

Thus, conditionally well-posed problems fall under the category of ill-posed. Their study has both theoretical and applied significance. Their investigation, as has already been mentioned, was started by A. N. Tikhonov who was followed by M. M. Lavrent'ev, V. K. Ivanov, and many of their disciples.

Numerous conditionally well-posed problems are associated with the solution of boundary-value problems for partial differential equations. In order to secure well-posedness of such problems, usually some additional requirements are imposed on the operators specifying the initial data. These requirements have been thoroughly studied for one group of boundary conditions. For parabolic equations they are called complementing condition and for elliptic and ordinary equations they are named Lopatinskii condition. Normally these conditions are written in terms of the transformed problem rather than the original problem and in addition, it is difficult to verify them. It would be, therefore, interesting to describe the condition of well- or ill-posedness in terms of the original problem, i. e. to find the coefficient condition. This important circumstance was first discovered by Godunov and Gordiyenko (1977).

The second group of questions is associated with the concept of solution itself. For comparatively long time, the researchers tried to find an exact (classical) solution of the ill-posed problem. However, it proved to be wrong for several reasons (Tikhonov and Arsenin, 1986). In many situations a principle for the selection of necessary solution was required satisfying supplementary information. Such information may be qualitative or quantitative.

The possibility to use supplementary quantitative information and the desire to overcome difficulties due to the absence of an ordinary solution of equation (1.1) brought Ivanov (1962) to the concept of a quasisolution generalizing the classical solution. Later, other generalizations were introduced, including "pseudosolution", "l-solution" etc. (Buhgeim, 1983; Morozov, 1987).

The next stage of the theory development is working out new efficient methods for the solution of ill-posed problems (Temirbulatov, 1989, 1996; Tikhonov, 1963, 1965; Tikhonov and Arsenin, 1986).

The above-mentioned set of questions will be treated in our study in varying degree of detail. They are presented in greater detail in the a more detailed presentation of these matters, which is given in the next section.

2. PROBLEMS AND METHODS FOR THEIR INVESTIGATION

2.1. Problems

We study the boundary-value problem of the form

$$\begin{aligned} \mathcal{E}(\partial)u(\alpha) &= 0 \qquad \text{in the domain } D; \quad u(\alpha) \in V, \\ \Gamma(\partial)u(\beta) &= \varphi \qquad \text{on the boundary } S = \partial D; \quad \varphi \in \Phi. \end{aligned} \tag{2.1}$$

Here, the differential operator $\mathcal{E}(\partial)$ is an equation (system), $\Gamma(\partial)$ specifies the boundary condition. The both operators with constant coefficients (elements), V and Φ are given sets (spaces) of functions including respectively the solution and the right-hand side of the boundary condition. The quadrant

$$D_{n+2} = \{t > 0,\ x > 0,\ y \in \mathbb{R}^n,\ n \geq 1\}$$

acts as the domain D.

The question is: are there many problems ill-posed in Hadamard sense among the problem of the type (2.1)?

The answer is presented in the form of a hypothesis.

Hypothesis. By fixing one of the operators $\mathcal{E}$ or Γ and varying the remaining one, we can always construct an ill-posed problem.

For the terms "fixing" and "varying" to be meaningful, differential equations and boundary data we conditionally subdivide into classes (groups). We can "fix" a specific operator or the whole class of those and by "varying" we understand a change in the operator inside a single class (horizontal change) and the transition from one class to another (vertical change).

By the equations classification we understand ODE and partial differential equations with the well-known typification.

We deem feasible to subdivide the problems in three groups depending on the boundary operator $\Gamma(\partial)$.

Group I includes main (standard) problems for the well-known types of differential equations. These problems are studied in the textbooks on equations and on the theory of boundary-value problems, there well-posedness is established using conventional methods: using the extremum principle and integral representation, a priori estimates, etc. Among those there are ill-posed problems, for example the Neumann problem for the Laplace equation is not always solvable, in Solonnikov (1965) a parabolic system is shown, for which the first boundary-value problem has a set of solutions.

Group II includes problems, for which the orders of derivatives in the boundary operator $\Gamma(\partial)$ are strictly less than the corresponding orders in the operator $\mathcal{E}(\partial)$ of the equation. It is for such problems that the complementing condition (Lopatinskii condition) defining the unique solvability makes sense, consequently, if it is violated, the problem is ill-posed.

We refer the problem to group III when no significant restriction is imposed on the construction and order of the operator $\Gamma(\partial)$. No consistent theory of such problems has been developed yet.

Thus, there arise at least two problems.

Problem 2.1. Construct ill-posed problems, i.e. single out ill-posed problems of the problems of the type (2.1).

Problem 2.2. Work out methods for research (solution) of ill-posed problems.

2.2. Methods

Partial differential equations are considered on the basis of the integral Fourier-Laplace transform and the notion of a nonclassical solution of a degenerate system of linear algebraic equations. The study is subdivided into two parts: direct approach and inverse approach.

2.2.1. Direct approach. Lowering the problem level

In the domain $D_{n+2} = \{t > 0,\ x > 0,\ y \in \mathbb{R}^n,\ n \geq 1\}$ consider the problem

$$\mathcal{L}(\partial_t, \partial_x, \partial_y)u(t,x,y) = f(t,x,y), \qquad l(\partial_t, \partial_x, \partial_y)u(t,x,y) = \varphi. \tag{2.2}$$

Let vector functions, i. e. the sought solution $u(t,x,y)$ and specified right-hand sides of $f(t,x,y)$, φ are such that integral Fourier or Laplace transforms can be applied to them. By their application, from (2.2) we arrive at a boundary-value problem for higher-order ODE on the ray $x \geq 0$

$$\begin{aligned} \mathcal{L}(p, d_x, i\omega)v(p,x,\omega) &= g(p,x,\omega), \qquad x > 0, \\ l(p, d_x, i\omega)v &= \psi(p,\xi), \qquad x = 0. \end{aligned} \tag{2.3}$$

Assume that (2.3) can be reduced to finding a bounded vector $y(x)$ on the ray $x \geq 0$ from the normal first-order system

$$\dot{y}(x) = A(p,\omega)y(x) + F, \qquad x > 0, \tag{2.4}$$

at $x = 0$ satisfying the boundary data

$$B(d_x)y(0) = b. \tag{2.5}$$

Further, consider a class of matrices A with two sorts of eigenvalues

$$\begin{aligned} &\operatorname{Re}\lambda_j(p,\xi) < -\delta, \quad j = 1,\dots,k;\\ &\operatorname{Re}\lambda_j(p,\xi) > \delta, \quad j = 1,\dots,m; \end{aligned} \tag{2.6}$$

with $k + m = N$, $1 \le k \le N$; $0 \le m < N$, $\delta > 0$.

The boundary matrix B is rectangular of dimension $k \times N$, $\operatorname{rang} B(1) = k$.

We denote by $\tilde{\Phi}(x)$ a part of the fundamental matrix of solutions of a homogeneous ($F \equiv 0$) system (2.4) corresponding to the first k eigenvalues λ_j and write its general bounded solution

$$y(x) = \tilde{\Phi}(x)C(p,\omega), \qquad C = (C_1, C_2, \dots, C_k). \tag{2.7}$$

By substituting the vector (2.7) into bounded equalities (2.5) we obtain the following system of linear algebraic equations (SLAE)

$$\mathcal{F}C = \tilde{b}, \qquad \mathcal{F} = \mathcal{F}(p,\omega) = B(d_x)\tilde{\Phi}(x)\big|_{x=0} \tag{2.8}$$

with respect to a "constant" vector.

The same algebraization can be performed in a different way. Let matrix $Z(x)$ have a set of linearly independent bounded vector solutions of the equations $\mathcal{L}(p, d_x, i\omega)v = 0$ as its columns. Then by substituting the general solution of the form

$$v(p, d_x, i\omega) = Z(x)C, \qquad C = (C_1, C_2, \dots, C_k) \tag{2.7$'$}$$

into the boundary data from (2.3) we again arrive at SLAE (2.8) where now

$$\mathcal{F} = l(p, d_x, i\omega)Z(x)\big|_{x=0}. \tag{2.8$'$}$$

Remark 2.1. If we allow for the relation between higher-order (2.3) and first-order (2.4) ODE systems we can write

$$\tilde{\Phi}(x) = \|Z(x), Z'(x), \dots, Z^{(r)}(x)\|^{\top},$$

where the number r is determined by the type and order of system (2.2).

Remark 2.2. The effect of system (2.4) nonhomogeneity can be easily transformed to boundary condition (2.5) by constructing Green's matrix.

2.2.2. The analogue of Lopatinskii condition of unique solvability

Therefore, well-posedness of a problem depends on the behavior of the matrix $\mathcal{F}(p, \omega)$ in the range of parameters of Laplace and Fourier transforms:

$$\begin{aligned} &\text{of the complex scalar} \quad p = p_0 + ip_1, \quad p_0 > 0, \quad p_1 \in \mathbb{R}, \\ &\text{and the real vector} \quad \omega \in \mathbb{R}, \quad -\infty < \omega_j < \infty, \quad j = 1, \ldots, n. \end{aligned} \tag{2.9}$$

This implies

I) If $\det \mathcal{F}(p, \omega) \neq 0$ for every such p and ω then Lopatinskii condition is true, which secures the unique solvability of the problem.

II) When $\det \mathcal{F}(p, \omega) \neq 0$, but if $|\det \mathcal{F} \leq \varepsilon$ is small, system (2.8) is ill-conditioned and the solution C no longer depends continuously on the vector $\tilde{b}$. This instability, sure enough, necessarily affects the solution of the previous problems.

III) If, on the other hand, $\det \mathcal{F} = 0$, even if for certain values of p and ω, it violates Lopatinskii condition and all the respective problems become ill-posed, i. e., not for every right-hand side they are solvable or have a whole family of solutions.

Therefore, using the direct approach problems of one of the following three levels were sold:

— initial problem (2.2) for partial differential equations;

— boundary-value problem (2.3) for parameterized system of ordinary differential equations;

— system of linear algebraic equations (2.8).

In summary, we can draw the following conclusions:

1) If the equality $\det \mathcal{F} = 0$ has the form of a manifold of the initial problem coefficients, it will be called a *domain* of ill-posedness.

2) The parameters p and ω_j making up the set P_0 of zeroes of the function $\det \mathcal{F}(p, \omega)$ can have the following features:

— they are completely free and assume any value from (2.9);

— part of them is fixed (defined) and the remaining part is arbitrary;

— the parameters are connected by a certain relation.

3) The conditions of ill-posedness (I-conditions) are of three types corresponding to the above-named features:

α) "coefficient condition" is the relation between the coefficients of the initial problem with free parameters p and ω_j;

β) "root" condition the parameters p and ω_j are connected;

γ) "coefficient and root condition", when part of the parameters is fixed, the remaining part is free, and there is a certain relation between the coefficients.

2.2.3. The inverse approach. The solution of the ill-posed problems of different levels

Further, the problems are solved when I-condition holds. This is done in the following order:

$$\text{SLAE} \Longrightarrow \text{ODE} \Longrightarrow \text{PDE}.$$

Step 1. Study SLAE (2.8) when its determinant is zero ($\Delta = \det \mathcal{F} = 0$). The purpose is to find the unique solution C^n (normal or pseudonormal Tikhonov and Arsenin (1986)). This can be achieved in different ways, some of which are listed below.

a) Let system (2.8) for the given vector $\tilde{b}$have the solution, which is obviously not unique, therefore by the definition of the normal solution we single out one solution $C^n(p, \omega)$.

b) Since on the set P_0 the [main] determinant of the system $\Delta = 0$, it will be required that all the determinants of the extended matrix should also be zero ($\Delta_j = 0$, $j = 1, \ldots, N$). Then the totality of independent solutions following from the last requirement leads to the solvability condition and in addition secures the existence of a family of solutions, out of which we again separate a normal one. On the other hand, by Kramer's formula we formally write $C_j = \Delta_j/\Delta$, $j \geq 1$, further by evaluating these uncertainties using L'Hopital's rule we find C^l.

b$'$) The influence of the set P_0 can be taken into account in the following way. Let us require that the set of zeroes N_j of the component $\tilde{b}_j$ should contain the given set P_0 ($P_0 \in N_j$) for every $j = 1, \ldots, N$.

c) If it is desirable to expand the set of right-hand sides of $\tilde{b}$ let us abolish the requirements $\Delta_j = 0$. In this case, we seek the pseudosolution of system (2.8) determined from the variational problem

$$J(C) = \|\mathcal{F}C - \tilde{b}\| \Longrightarrow \min_{C_j}.$$

The resulting new SLAE can also prove degenerate thus rendering the process cyclical. However, to complete the cycle let us find the unique pseudosolution C^p.

Let us summarize:

— in all the above-mentioned situations the solvability condition ([possibly] necessary) is written that restricts the class of the right-hand sides;

— irrespective of the form, in which the I-condition is specified, the parameters p and ω_j, generally speaking,become free in all the calculations except those involving boundary conditions and solvability condition;

— $C^n(p,\omega)$ is an ordinary (exact) solution of system (2.8) because it is singled out of the classical solutions;

— $C^l(p,\omega)$ is the approximated solution of SLAE (2.8);

— for $C^p(p,\omega)$ not all equalities from (2.8) are true.

Finally, here are some typical forms of the coefficient $C(p,\omega)$:

α) the function $\tilde{f}(p,\omega)$ is specified by independent arguments;

β) $f(\omega) = \tilde{f}(p,\omega)|_{p=\varphi(\omega)}$, here the function $\varphi(\omega)$ is strictly positive at the same time there can hold the restrictions of the type $\omega \neq 0$ or $\omega_j > 0$ (perhaps, in some other form).

γ) $f(p) = \tilde{f}(p,\omega)|_{\omega_j=\omega_{0j}}$, where ω_{0j} is fixed (in particular, $\omega_{0j} = 0$).

Step 2. The solution of problem (2.4), (2.5) or (2.3) for ODE is reconstructed according to the representation (2.7) or (2.7$'$). Here, two solution types can arise:

— ordinary, when all the equalities of system (2.4) and boundary data (2.5) (or all equalities (2.3)) are strictly satisfied (e. g., for the case α);

— nonclassical satisfying the system of ODE, but at the same time not all the boundary equalities are true or these equalities are not universally present (in the cases β and γ).

Step 3. Application of inverse integral transform on ODE solution (with respect to functions $v(p, x, \omega)$), generally speaking, follows the standard procedure. And when the solvability condition is transformed and the original of the coefficient $C(p, \omega)$ is calculated (with the exception of the coefficient case α) an specially arranged method is suggested. First we reverse the transform with free parameters, and then substitute the values of the remaining parameters (fixed in the γ case or connected in the β case) into the integrand of the remaining transforms. The result of the γ will be left unchanged and for the β case further inversion may be necessary.

The main thing is that such "solution" exists in the domain of ill-posedness (when the solvability condition is true), it is unique and the solution can be calculated explicitly.

Therefore, we came back to the solution of the initial problem (2.2), which now has become conditionally well-posed.

3. DENOTATION AND TERMS

In order to render presentation and statement formulation simpler, we will use the well-known denotations: differentiation with respect to x_k and to t

$$\partial_{x_k} = \frac{\partial}{\partial_{x_k}}, \qquad k = 1, 2, \ldots, n; \qquad \partial_t^s = \frac{\partial^s}{\partial t^s};$$

Δ is the Laplace operator, Δ_y is the Laplace operator with respect to y; I is the unit operator, and $I^{(N)}$ is the unit matrix of the order $N \times N$.

Let $k = (k_1, k_2, \ldots, k_n)$ be an arbitrary vector with n nonnegative components and $\xi = (\xi_1, \xi_2, \ldots, \xi_\nu)$ be an arbitrary n-dimensional vector, then $\xi^k = (\xi_1^{k_1}, \xi_2^{k_2}, \ldots, \xi_n^{k_n})$ and at the same time the components of the vector ξ may be numbers or operators, in particular, for $\xi = \partial_x^k$, the symbol ∂_x^k stands for the differential operator

$$\partial^k = \partial_{x_1}^{k_1} \partial_{x_2}^{k_2} \ldots \partial_{x_n}^{k_n}$$

of the order

$$k = k_1 + k_2 + \cdots + k_n.$$

The direct Fourier transform over the variable x is denoted by F_x^+, and the inverse F_ξ^-:

$$u(t, x) \xrightarrow{\mathrm{F}_x^+} \tilde{u}(t, \xi), \qquad \tilde{u}(t, \xi) \xrightarrow{\mathrm{F}_\xi^-} u(t, x).$$

The Laplace transform over the time t is designated by L_t^+, and the inverse transform is designated by L_p^-:

$$u(t,x) \xrightarrow{\mathrm{L}_t^+} \bar{u}(p,x), \qquad \bar{u}(p,x) \xrightarrow{\mathrm{L}_p^-} u(t,x).$$

The result of the successful application of those transforms is expressed as $\hat{u}(p,\xi) \equiv V(p,\xi)$.

4. SOME FACTS FROM THE THEORY OF MATRICES

4.1. Notation

— $A = (a_{ij})$, $i,j = 1,\dots,N$ is a square matrix of the order N with constant elements; its norm $\|A\|^2 = \sum_{i,j=1}^N a_{i,j}^2$;

— $\Delta(\lambda) = \det|A - \lambda I|$ is the characteristic polynomial of the matrix , or, to put it more precisely, the determinant of the characteristic matrix $A(\lambda) = (A - \lambda I)$;

— $B(\lambda) = \|b_{ik}(\lambda)\|_1^N$is the adjoint matrix, $b_{ik}(\lambda)$ is the algebraic complement of the element $(a_{ki} - \lambda\delta_{ik})$ in $\Delta(\lambda)$;

— $d_{n-1}(\lambda)$ is the greatest common divisor of all the elements of the matrix $B(\lambda)$;

— (λ) is the "reduced" adjoint matrix for $A(\lambda)$. There holds an equality $B(\lambda) = d_{n-1}(\lambda)C(\lambda)$;

— $\Psi(\lambda)$ is the minimum multinomial of the matrix (annihilating multinomial $\equiv \psi() = 0$ of the lowest degree with the leading coefficient equal to unity, i. e. reduced multinomial);

— r_j is the rank of the characteristic matrix $A(\lambda_j)$;

— k_j is algebraic or [simple] multiplicity of the eigenvalue λ_j;

— $k_j^g = N - r_j$ is geometric multiplicity of the eigenvalue λ_j (the maximum number of linearly independent eigenvectors related to λ_j). One eigenvector cannot correspond to two different eigenvalues;

— m_j^P is the number of main adjoint vectors, which correspond to λ_j $(k_j^g + m_j^P = k_j)$;

— $m = \sup_j (N - r_j)$ is the number of cells in the Frobenius form.

4.2. Formulas

Let us write out the characteristic polynomial in the form

$$\Delta(\lambda) = \lambda^N - p_1\lambda^{N-1} - \cdots - p_{N-1}\lambda - p_N. \tag{4.1}$$

By introducing a multinomial for $\lambda \neq \mu$ (numbers)

$$\delta(\lambda,\mu) = \frac{\Delta(\lambda)-\Delta(\mu)}{\lambda-\mu} = \lambda^{N-1} + (\mu - p_1)\lambda^{N-2} + (\mu^2 - p_1\mu - p_2)\lambda^{N-3} + \ldots$$

we can find the matrix

$$B(\lambda) = \delta(\lambda I, A) = \lambda^{N-1}I + B_1\lambda^{N-2} + \cdots + B_{N-1}. \tag{4.2}$$

The coefficients p_k and B_k from formulas (4.1) and (4.2) are calculated by the following recurrence relations

$$\begin{gathered} A_1 = A, \qquad P_k = K^{-1}\operatorname{Sp} A_k, \qquad B_k = A_k - P_k I, \\ A_k = AB_{k-1}, \qquad K = 1,\ldots,N. \end{gathered} \tag{4.3}$$

(D. K. Fadeev's method) or

$$B_0 = I, \quad B_k = AB_{k-1} - P_k I, \quad k = 1,\ldots,N-1 \tag{4.4}$$

(Gantmacher's method).

From the control equality $AB_{N-1} = P_N I$ (present in the both methods) the inverse matrix can be found

$$A^{-1} = B_{N-1}/P_N.$$

Remark 4.1. Any nonzero column of $B(\lambda_j)$ (such a column exists) determines eigenvectors, corresponding to eigenvalue λ_j.

4.3. Definitions

Further, let there hold the expansion

$$\Delta(\lambda) = (\lambda-\lambda_1)^{k_1}(\lambda-\lambda_2)^{k_2}\ldots(\lambda-\lambda_r)^{k_r}, \qquad \sum_{j=1}^{r} k_j = N, \tag{4.5}$$

here $\lambda_1,\ldots,\lambda_r$ are pairwise different.

Let us use the notation $\varphi_{ij}(\lambda) = (\lambda - \lambda_{ij})^{k_j}$, then the vector space

$$X = \ker \varphi_1(A) + \cdots + \ker \varphi_r(A),$$

where $\ker \varphi_j(A)$ is the root subspace of the matrix A, corresponding to eigenvalue λ_j and its dimensionality equals the algebraic multiplicity of this eigenvalue λ_j.

Definition 4.1. Vectors of root subspaces are called *root* vectors. The least integer s, for which $(A - \lambda_j I)^s x = 0$ determines the *height* of the root vector x.

Then the system of vectors

$$y_s = x, \quad y_{s-1} = (A - \lambda_j I)x, \quad \ldots, \quad y_1 = (A - \lambda_j I)^{s-1} x \tag{4.6}$$

will be linearly independent.

The equalities $Ay_1 = \lambda_j y_1$, $Ay_2 = \lambda_j y_2 + y_1$, $\ldots$, $Ay_s = \lambda_j y_s + y_{s-1}$ imply that the root vectors of the height s comprise one eigenvalue y_1 and adjoint vectors $y_2, y_3, \ldots, y_s$. Therefore a root vector of a unity height is an eigenvector.

Now we number vectors from (4.6) in the inverse order

$$z_1 = y_s = x, \qquad z_2 = y_{s-1} = (A - \lambda_j I)x, \quad \ldots,$$
$$z_s = y_1 = (A - \lambda_j I)x, \tag{4.6'}$$

then $Az_1 = \lambda_j z_1 + z_2$, $Az_2 = \lambda_j z_2 + z_3$, $\ldots$, $Az_s = \lambda_j z_s$.

Obviously, vectors $(z_K)_1^s$ are also linearly independent and correspond to eigenvalue λ_j, or, to be more precise, to the factor $(\lambda - \lambda_j)^{k_j}$.

Thus, let $s_1, \ldots, s_r$ be the maximum heights of the root vectors corresponding to pairwise disjoint eigenvalues: $\lambda_1, \lambda_2, \ldots, \lambda_r$. Then the following equality holds true

$$\Psi(\lambda) = (\lambda - \lambda_1)^{s_1}(\lambda - \lambda_2)^{s_2} \ldots (\lambda - \lambda_r)^{s_r}, \qquad s_j \leq k_j, \quad j \geq 1. \tag{4.7}$$

Let us assume that the characteristic matrix has the rank r. Let us denote by $d_j(\lambda)$, $j \leq r$, the reduced greatest common divisor of all minors of the j-th order of the matrix $A(\lambda)$, at the same time it is assumed that $d_0(\lambda) = 1$. The multinomial $d_j(\lambda)$ is divisible by $d_{j-1}(\lambda)$ for every $j \leq r$. The following formulas hold true:

$$\psi(\lambda) = \Delta(\lambda)/d_{n-1}(\lambda),$$
$$B(\lambda) = d_{n-1}(\lambda)C(\lambda), \qquad C(\lambda) = \Psi(\lambda I, A) \tag{4.8}$$

where $\Psi(\lambda, \mu) = (\psi(\lambda) - \psi(\mu))/(\lambda - \mu)$, $\lambda \neq \mu$.

Every nonzero column of the matrix $C(\lambda_j)$ determines eigenvector, corresponding to eigenvalue λ_j.

Definition 4.2. *Invariant multinomials* of the characteristic matrix of rank r is the term used for the reduced polynomials:

$$I_1(\lambda) = \frac{d_r(\lambda)}{d_{r-1}^{(\lambda)}}, \quad I_2(\lambda) = \frac{d_{r-1}(\lambda)}{d_{r-2}^{(\lambda)}}, \quad \dots, \quad I_r(\lambda) = \frac{d_1(\lambda)}{d_0(\lambda)}, \quad d_0 = 1. \quad (4.9)$$

Each of them is divisible by the subsequent one. These multinomial at $r = N$ are invariant multinomials of the characteristic matrix $A(\lambda)$ and there holds the equality

$$\Delta(\lambda) = d_N(\lambda) = I_1(\lambda) I_2(\lambda) \dots I_N(\lambda). \quad (4.10)$$

Let us represent $I_j(\lambda)$ by canonical expansions

$$I_j(\lambda) = (\lambda - \lambda_1)^{\alpha_{j1}} (\lambda - \lambda_2)^{\alpha_{j2}} \dots (\lambda - \lambda_S)^{\alpha_{js}}$$
$$(\lambda_j \neq \lambda_k, \ j \neq k), \qquad j = 1, \dots, N, \quad (4.11)$$

where the following relations hold true

$$0 \le \alpha_{Nm} \le \dots \le \alpha_{2m} \le \alpha_{1m} \le k_m, \qquad \sum_{j=1}^{N} \alpha_{jm} = k_m \quad (4.12)$$

for every $1 \le m \le N$. Some of these numbers can be zeroes.

Definition 4.3. Every factor $(\lambda - \lambda_m)^{\alpha_{jm}}$ entering in the expansions (4.11) at $\alpha_{jm} > 0$, is called an *elementary divisor* of the matrix A.

For more details concerning theory of matrices see Bellman (1970), Voyevodin (1974), Voyevodin and Kuznetsov (1984). Gantmacher (1966), Lancaster (1978), Smirnov (1953, 1957).

5. CANONICAL (NORMAL) FORMS OF MATRICES

5.1. Direct problem

We reduce the given matrix A of the order N to the canonical form.

Smith form. Every λ-matrix of the rank $r \le N$ is equivalent to the diagonal matrix $I(\lambda)$, which is called Smith's form and compiled of invariant multinomials

$$I(\lambda) = \operatorname{diag}(I_1(\lambda), I_2(\lambda)_3, \dots, I_r(\lambda), 0, \dots, 0). \quad (5.1)$$

Since the invariant multinomial form a complete system of invariants λ-matrix, the Smith's form, or, to be more precise, multinomials $I_j(\lambda)$, $j = 1, \ldots, r$, are uniquely defined by the assignment of the λ-matrix. From form (5.1), according to the notations and relations (4.9), (4.12) we find

$$\psi(\lambda) = I_1(\lambda) \quad \text{and} \quad \Delta(\lambda) = \prod_{j=1}^{N} I_j(\lambda). \tag{5.2}$$

Frobenius form. It is always possible to select a (transforming) non-degenerate the matrix T so that $P = T^{-1}AT$ should have the first natural normal form, which is called Frobenius form:

$$P_I = \operatorname{diag}\{P_1, P_2, \ldots, P_m\}. \tag{5.3.I}$$

The Frobenius matrix

$$P_j = \begin{bmatrix} 0 & 0 & \ldots & -p_{j,s_j} \\ & & & -p_{j,s_j-1} \\ & (I^{s_j-1}) & & \ldots \\ & & & -p_{j,1} \end{bmatrix}, \qquad \sum_{j=1}^{m} s_j = N \tag{5.4}$$

is accompanied by the characteristic polynomial

$$q_j(\lambda) = \lambda^{s_j} + p_{j,1}\lambda^{s_j-1} + \cdots + p_{j,s_j-1}\lambda + p_{j,s_j}. \tag{5.5}$$

Let us introduce the notation

$$\chi_1(\lambda), \ \chi_2(\lambda), \ \ldots, \ \chi_q(\lambda) \tag{5.6}$$

for elementary divisors of the matrix A in the number field K and let them be accompanied by the matrices $P^{(j)}$, $j = 1, \ldots, q$. Since $\chi_j(\lambda)$ is the only elementary divisor as well as the only invariant multinomial of the matrix $P^{(j)}$, $j = 1, \ldots q$, the quasidiagonal matrix

$$P_{\mathrm{II}} = \{P^{(1)}, P^{(2)}, \ldots, P^{(q)}\} \tag{5.3.II}$$

as elementary divisor has multinomials $\chi_j(\lambda)$.

Thereby, the matrices A and P_{II} have the same elementary divisors in the field K and there exists matrix W ($|W| \neq 0$) such that $A = WP_{\mathrm{II}}W^{-1}$. Here the matrix P_{II} is called the second natural normal form. The first and the second natural normal forms are characterized by the following properties:

— quasidiagonal form (5.3.I) and (5.3.II);

— special structure of diagonal structure of diagonal cells (5.4);

— characteristic and minimum polynomial (5.5) corresponding to each cell (in the case of $P_{\rm II}$ it is the degree of the multinomial nonreducible in the field K);

— each characteristic multinomial $q_j(\lambda)$, $j = 1, \dots, m$ in the series starting from the second one will be divisor of the previous one;

— invariant multinomials of the matrix P_j are equal to $\{q_j(\lambda), 1, \dots, 1\}$, and therefore the series (4.9) of the characteristic multinomial is the invariant of the multinomials of the matrix P, consequently, it will be the parent matrix A;

— if λ_{0j} is the root of the polynomial $q_j(\lambda)$, then irrespective of its multiplicity there exists one eigenvector of the form $\{1, \lambda_{0j}, \lambda_{0j}^2, \dots, \lambda_{0j}^{S_j-1}\}$.

From the above-mentioned properties of form (5.3.I) there follow the equalities

$$\Psi(\lambda) = q_1(\lambda), \qquad \Delta(\lambda) = q_1(\lambda) \dots q_m(\lambda). \tag{5.2$'$}$$

Remark 5.1. Elementary divisors of the matrix A unlike the invariant polynomials are substantially connected with the given number field K. If instead the initial field we take a different field (also comprising elements of the given matrix), then elementary divisors may be different. For example, let A be a matrix with real-valued elements, evidently, its characteristic multinomial will have real coefficients, and among the roots there can be complex roots. Let us assume that K is a field of real numbers, then there will be degrees of nonreducible square trinomials among other elementary divisors, and if K is a field of complex numbers, each elementary divisor has the simplest form $(\lambda - \lambda_0)$.

Jordan form. Now the set of elementary divisors (5.4) for every eigenvalue will be rewritten in the form

$$(\lambda - \lambda_1)^{p_1},\ (\lambda - \lambda_2)^{p_2},\ \dots,\ (\lambda - \lambda_S)^{p_S}, \qquad \sum_{j=1}^{s} p_j = N, \quad p_j \geq 1. \tag{5.6$'$}$$

Some of the numbers λ_k, $k = 1, \dots, s$, may be the same.

Theorem 5.1. *In order to reduce the given matrix* A *to the diagonal form it is necessary and sufficient that* $r_j = N - k_j$ *for every eigenvalue* λ_j.

Theorem 5.2. *If* A *has an elementary divisors of form* (5.6′), *then there exists a nondegenerate matrix* V *such that* $V^{-1}AV$ *will have a quasidiagonal (or diagonal) form, which is called Jordan form*

$$V^{-1}AV = J = \{J_{P_1}(\lambda_1), J_{P_2}(\lambda_2), \dots, J_{P_S}(\lambda_S)\}. \tag{5.7}$$

The matrix $J_P(\lambda_0)$ is called a Jordan block (cell)

$$J_p(\lambda_0) = \begin{bmatrix} \lambda_0 & 1 & 0 & \dots & 0 & 0 \\ 0 & \lambda_0 & 1 & \dots & 0 & 0 \\ 0 & 0 & \lambda_0 & \dots & 0 & 0 \\ \dots & \dots & \dots & \dots & \dots & \dots \\ 0 & 0 & 0 & \dots & \lambda_0 & 1 \\ 0 & 0 & 0 & \dots & 0 & \lambda_0 \end{bmatrix} \quad \text{or} \quad J_p(\lambda_0) = \begin{bmatrix} \lambda_0 & 0 & 0 & \dots & 0 & 0 \\ 1 & \lambda_0 & 0 & \dots & 0 & 0 \\ 0 & 1 & \lambda_0 & \dots & 0 & 0 \\ \dots & \dots & \dots & \dots & \dots & \dots \\ 0 & 0 & 0 & \dots & \lambda_0 & 0 \\ 0 & 0 & 0 & \dots & 1 & \lambda_0 \end{bmatrix}. \tag{5.8}$$

The subscript p and argument λ_0 respectively show the order of the matrix and the number on the main diagonal. If $p = 1$, then $J(\lambda_0) = \lambda_0$ is simply a number. It can be readily verified that Jordan block (5.8) has only one elementary divisor of the form $(\lambda - \lambda_0)^p$ and this block is similar to the Frobenius matrix for the polynomial $q(\lambda) = (\lambda - \lambda_0)^p$. To the accuracy of Jordan blocks permutation every matrix can be reduced to the unique (5.8).

This form has the following characteristics:

— quasidiagonal form (5.7);

— special structure of diagonal cells (5.8);

— characteristic polynomial of every cell is a degree of a multinomial irreducible in the field K;

— viewed in the order of (4.6) or (4.6′) vectors form columns of the transforming matrix V, in blocks $J_p(\lambda_0)$ there are “units” above or below the main diagonal.

Thus along the main diagonal of blocks $J_p(\lambda_0)$ there are eigenvalues then the transforming matrix V consists of eigenvectors and adjoined vectors ($AV = VJ$). Consequently, all the eigenvalues and elementary divisors are readily found from Jordan form.

5.2. Relation between the forms. Uniqueness of the direct problem solution

If the aggregate of elementary divisors is known, then all types of canonical forms can be written. Indeed:

— invariant multinomials from Smith form are directly calculated by formula (4.11);

— characteristic polynomials $g_j(\lambda)$ of the Frobenius matrix P_j are formed from elementary divisors as $g_j(\lambda) = (\lambda - \lambda_j)^{P_j}$;

— each elementary divisors shows the order and diagonal element of Jordan block.

As has already been noted, any of these forms is uniquely defined, which means that a direct problem of the canonical form (matrix reducing to its canonical form) has one solution.

5.3. Notes on algorithms

Smith form can be obtained by so-called elementary operations on matrices. The well-known methods suggested by V. I. Krylov and A. M. Danilevskii (intended for the transformation of the secular equation) reduce the matrix $A(\lambda)$ to the first normal form. Algebraic analysis of V. I. Krylov's method is presented in Gantmacher (1966) where different ways of constructing the matrix V are shown, and in the works by Smirnov (1953) a universal method for the problem solution is given briefly retold in Temirbulatov (1996).

The simplest and best illustrative approaches to obtaining the matrix T are shown in the book by Godunov (1986).

5.4. A method for constructing matrix V

A. Let for the matrix A of the order N the total set of elementary divisors (5.6) be known and the reduced matrix $C(\lambda)$. After substituting the minimum multinomial $\psi(\lambda)$ in the form

$$\psi(\lambda) = (\lambda - \lambda_0)^m \chi(\lambda), \qquad \chi(\lambda_0) \neq 0, \tag{5.9}$$

let us [assign] to it the following set of m matrices

$$C = C(\lambda_0), \quad D = \frac{C'(\lambda_0)}{1!}, \quad F = \frac{C''(\lambda_0)}{2!}, \quad \ldots, \quad G = \frac{C^{(m-1)}(\lambda_0)}{(m-1)!}. \tag{5.10}$$

Proceeding from the definition of the matrix $C(\lambda)$ it is always possible to select a nonzero column $C_k \neq 0$ and then m columns corresponding to it

$$C_k, \; D_k, \; \ldots, \; Q_k, \; G_k \tag{5.11}$$

will be linearly independent vectors forming a Jordan chain of columns.

There hold the following basic formulas

$$\begin{cases} (-\lambda_0 I)C_k = 0, \\ (A-\lambda_0 I)D_k = C_k, \\ \cdots\cdots\cdots\cdots\cdots\cdots \\ (A-\lambda_0 I)G_k = Q_k, \end{cases} \qquad k = 1, 2, \ldots, N, \tag{5.12}$$

showing that C_k is eigenvector, and the rest are adjoined vectors.

If $C_{\bar{k}} = 0$ for some $k = \bar{k}$, but $D_{\bar{k}} \neq 0$, then columns $D_{\bar{k}}, F_{\bar{k}}, \ldots, G_{\bar{k}}$ form a Jordan chain of $(m-1)$ vectors.

B. Let us first consider the case when elementary divisors (5.6′) are mutually disjoint, i. e., in them $\lambda_i \neq \lambda_j$, $i \neq j$, $i, j = 1, \ldots, s$. To each elementary divisor of the form $(\lambda - \lambda_k)^{p_k}$, $k = 1, \ldots, s$ we assign its own Jordan chain of columns of form (5.11). Then the matrix V is determined as follows

$$V = \{C_1, D_1, \ldots, G_1; \; C_2, D_2, \ldots, G_2; \; \ldots; \; C_s, D_s, \ldots, G_s\}. \tag{5.11.1}$$

Obviously, $\det V \neq 0$.

Remark 5.2. The following transformations of the matrix V are allowed:

i) multiplication of all the columns of some Jordan chain by a nonzero number;

ii) addition to each (starting from the second) Jordan chain column of the previous column of the same chain, previously multiplied by the same arbitrary number;

iii) addition to all the Jordan chain columns of the respective columns of another chain consisting of the same or greater number of columns and corresponding to the same eigenvalue λ_0.

Clearly, all these transformations enable us to simplify the matrix V.

C. General case. Let us assume that to the number λ_0 there correspond elementary divisor of the form $(\lambda - \lambda_0)^m$ whose number is p; elementary divisor of the form $(\lambda - \lambda_0)^{m-1}$ whose number is q; divisors $(\lambda - \lambda_0)^{m-2}$ whose number is r, and so on.

Let us first note some properties of matrices (5.10)

α) they have respective ranks: p, $2p+q$, $3p+2q+r$, $\dots$;

β) column of each of them is a linear combination of columns from any successive matrix;

γ) they may be represented as multinomials of A:

$$C = h_1(A), \quad D = h_2(A), \quad \dots, \quad G = h_m(A) \tag{5.13}$$

and at the same time

$$h_k(\lambda) = \Psi(\lambda)/(\lambda - \lambda_0)^k, \quad k = 1, \dots, m. \tag{5.14}$$

δ) without changing basic formulas (5.12) in C any column can be replaced by an arbitrary linear combination of all the columns after the corresponding substitution in $D, \dots, G$.

Thus, let us write a Jordan chain for the general variant

$$\left\{\begin{array}{l} \underbrace{\overbrace{(C_1, D_1, \dots, G_1)}^{(m)}, \dots, \overbrace{(C_p, D_p, \dots, G_p)}^{(m)}}_{p} \\ \underbrace{\overbrace{(D_{2p+1}, \dots, G_{2p+1})}^{(m-1)}, \dots, \overbrace{(D_{2p+q}, \dots, G_{2p+q})}^{(m-1)}}_{q} \end{array}\right. \tag{5.15}$$

Let matrix A further have s different eigenvalues λ_j, $j = 1, \dots, s$, and let multinomials Δ and Ψ be represented in the form $\Delta(\lambda) = \prod_j^s (\lambda - \lambda_j)^{p_j}$ and $\Psi(\lambda) = \prod_{j=1}^s (\lambda - \lambda_j)^{m_j}$, $m_j \le p_j$.

For every number λ_j let us form its own system of independent Jordan chains (5.15). The number of columns in this system will be equal to p_j and consequently, the total set of chains contains $N = \sum_{j=1}^s p_j$ columns, which form one of the desired transforming matrices V.

5.5. Semi-inverse problem

Let us write canonical form of the given matrix without specifying the transforming matrices T or V. It is sufficient to know the total set of elementary divisors.

There is a special case, for which the desired forms can be written in a very simple way. It is when the Frobenius form contains one matrix P_{I}. In this case we have

$$\Delta(\lambda) = \Psi(\lambda) = g_1(\lambda) = (\lambda - \lambda_1)^{k_1}(\lambda - \lambda_2)^{k_2} \dots (\lambda - \lambda_r)^{k_r},$$
$$\sum_{j=1}^{r} k_j = N. \tag{4.5}$$

Hence, according to the properties of the characterizing forms, we write $I(\lambda) = \operatorname{diag}\{g_1(\lambda), 1, \dots, 1\}$, $J = \operatorname{diag}\{J_{k_1}(\lambda_1), \dots, J_{k_r}(\lambda_r)\}$, and matrix $P = P_1$ is written proceeding from the coefficients of the polynomial $g_1(\lambda)$.

Let us return to the general variant, i. e. to the construction of the elementary divisors corresponding to eigenvalue λ_0. Now we show several simple ways to do it

A. For small-size matrix A it is better directly to find eigenvectors and adjoined vectors. Then from their quantity it is easy to obtain elementary divisor and polynomial $\Psi(\lambda)$.

B. Obviously, in the general case eigenvalue λ_0 can correspond to the elementary divisor of the form

$$\{(\lambda - \lambda_0), \dots, (\lambda - \lambda_0); (\lambda - \lambda_0)^2, \dots, (\lambda - \lambda_0)^2; \dots; (\lambda - \lambda_0)^m, \dots (\lambda - \lambda_0)^m\}.$$

Let us denote their quantity in every group by the number n_j, whereas $n_j \geq 0$ for $j = 1, \dots, (m-1)$, $n_m > 0$. The greatest index m is calculated as the index at which $\delta_{m-1} < \delta_m = \delta_{m+1}$, it should be recalled that δ_k is the defect of the matrix $(A - \lambda_0 I)^k$. Now the numbers n_j of interest for us are calculated using the equality

$$n_j = -\delta_{j-1} + 2\delta_j - \delta_{j+1},$$
$$\delta_0 = 0, \qquad \delta_{m+1} = \delta_{mj}, \quad j = 1, \dots, m. \tag{5.16}$$

5.6. Inverse problem

Let us reconstruct the matrix A from its eigenvalues or from its canonical form. To this end, let one of the following assumptions hold true:

α) the total set of eigenvalues $\{\lambda\}$ is given;

β) part of eigenvalues λ_j, $\operatorname{Re}\lambda_j < 0$, $j = 1, \dots, k$ is given, and, in addition, it is known that the other part of eigenvalues λ_j, $j = 1, \dots, m$, is such that $\operatorname{Re}\lambda_j > 0$ $(k + m = N,\ 1 \le k \le N)$;

γ) a certain canonical form (for example, Jordan form) is given.

Clearly, the problem solution is a whole family, because the class of such matrices has one and the same set of eigenvalues (or one and the same canonical form).

Solution. It is sufficient to reduces the cases α and β to the case γ or, to be more precise, to write some canonical form and multiply the latter by any (transforming) matrix, as required.

6. ON SOLUTIONS OF SYSTEM OF ALGEBRAIC EQUATIONS

Let us consider a system of the general form (Tikhonov and Arsenin, 1986)

$$Ax = b \tag{6.1}$$

with a rectangular matrix A of the dimensionality $m \times n$. Let us assume that the number of equations is m, number of unknowns is n and the rank of the matrix of the system is r, all of them are absolutely arbitrary. First recall the well-known definition of exact solution.

Definition 6.1. The classical (exact) solution of equation (6.1) is such an element $\bar{x} \in \mathbb{R}^n$, at which $\|A\bar{x} - b\|_{\mathbb{R}^n} = 0$.

Then, without loss of generality we can consider that the principal determinant of the system is located at the upper left corner:

$$\Delta_r \neq 0, \qquad \Delta_r = |a_{ij}|_{i,j=1}^{r}.$$

Let us form $(m - r)$ determinants of the order $r + 1$, called characteristic and formed by adding to the principal determinant one row containing coefficients of the equations with the number greater than r, and column of absolute terms.

Consider the problem of one-valued solvability of this system.

Theorem 6.1. *System* (6.1) *is solvable if and only if all the characteristic determinants of the system are zero.*

Or in other words

Theorem 6.1′ [Kronecker—Capelli]. *For system* (6.1) *to be compatible, it is necessary and sufficient that the rank of the extended matrix of the system should be equal to the rank of the matrix A, i. e. to the quantity r.*

Let the system be solvable, then its solution is a whole family, when $r < N$, which we will write as

$$x_j = \alpha_j + \beta_{k+1}^{(j)} x_{r+1} + \cdots + \beta_r^{(j)} x_n, \qquad j = 1, \ldots, r, \tag{6.2}$$

where α_j, β_k are some constants.

Clearly, the general solution of the reduced homogeneous system forms a subspace of the dimensionality $m - r$ in $\mathbb{R}^m$ and any its basis is called *fundamental system of solutions* of a homogeneous system.

Theorem 6.2. *System* (6.1) *has one solution, if and only if it is compatible and moreover, $r = n$.*

Let $m = n$, then the system

$$\bar{A}^\top y = 0 \tag{6.3}$$

is called union or conjugate homogeneous. It is obtained from (6.1) by replacing rows with columns and all the numbers and by conjugates.

Theorem 6.3. *If* $\det A = 0$, *then for the solution of system* (6.1) *to exist, it is necessary and sufficient, that the right-hand side b should be orthogonal to all the vectors being a solution of the union system* (6.3)*:* $(b, y_j) = 0$.

6.1. A case of one-valued solvability violation

Classical solution can fail to exist or be non-unique. To overcome these difficulties new solution types are introduced. Let system (6.1) be degenerate and at the same time let the vector of the right-hand side satisfy the conditions of solvability (from Theorems 6.1, 6.1′ and 6.3). Then the solution is non-unique, and let X_A be a total set of all its solutions. In this case we determine the normal solution.

Definition 6.2. The vector x_n is called the normal solution of system (6.1) with respect to the fixed vector x^1 if

$$\inf_{x \in X_A} \|x - x^1\|_{\mathbb{R}^n} = \|x_H - x^1\|. \tag{6.4}$$

Usually x^1 is obtained from the additional information about solution. If there is no such information, it is assumed that $x^1 = 0$ (zero). The main thing is that the determined normal solution is unique. That is: if system (6.1) is compatible, then among its solutions there exist only one belonging to $\operatorname{im} A^*$ and called normal. How can it be calculated?

Let the vectors $\bar{x}_{r+1}, \bar{x}_{r+2}, \dots, \bar{x}_n$ form the basis of the linear space $N_A \equiv \ker A$. The solution $\bar{x}^0$ of system (6.1) which satisfies $(n - r)$ conditions conditions of orthogonality $(\bar{x}^0 - x^1, \bar{x}_s) = 0$, $s = r + 1, \dots, n$ is uniquely determined, and it will be a normal solution.

When there is no classical solution, a pseudosolution is sought.

Definition 6.3. The element $\hat{x}$ is called the pseudosolution of system (6.1) if it minimize the error of the vector x:

$$\min_{x \in \mathbb{R}^n} \|Ax - b\|_{\mathbb{R}^m} = \|A\hat{x} - b\|. \tag{6.5}$$

Obviously, if $\|A\hat{x} - b\| = 0$, pseudosolution coincides with the classical solution $\bar{x}$, thus pseudosolution generalizes the notion of the exact solution. In particular, pseudosolution can be obtained as a solution of the system $A^*Ax = A^*b$.

Equation (6.1) can have a set of pseudosolutions. In this case normal pseudosolution is found in the obvious way. Thus, for every system of linear algebraic equations there always exists normal pseudosolution, which is also unique. Among all the pseudosolutions of the system only normal pseudosolution is orthogonal to $\ker A$, i. e. $\hat{x}_n \perp \ker A$.

Let us present one more method for calculation of the element $\hat{x}_n$.

Definition 6.4. The matrix A^+ of dimension $(n \times m)$ is called pseudoinverse for A of dimension $(m \times n)$, if the following conditions are fulfilled

$$AA^+A = A, \qquad A^+ = UA^* = A^*V \qquad (A^+ = A^*WA^*),$$

where U, W, V are some nondegenerate matrices.

Now normal pseudosolution $\hat{x}_n$ every system of linear algebraic equations can be found from the equality

$$\hat{x}_n = A^+b.$$

Remark 6.1. 1) Pseudoinverse matrix A^+ is the best approximate solution (by least-squares method) of the matrix equation $Ax = I$, i.e. $A^+ = (A'A)^{-1}A'$. 2) The matrix A^+ can also be calculated using Greville's method (Gantmacher, (1966).

7. RELATION BETWEEN AN ARBITRARY EQUATION AND A SYSTEM OF FIRST-ORDER ORDINARY DIFFERENTIAL EQUATIONS. GREEN'S MATRIX

Consider linear equations

$$\dot{y} = A(t) + \vec{b}(t), \qquad t \in (a_1, a_2), \tag{7.1}$$

where $y(t)$, $\vec{b}(t)$ are vectors with N components, $A(t)$ is a square matrix (Godunov, 1994; Pontryagin, 1988). All the components of vectors and elements of the matrix are continuous over the interval (a_1, a_2), it is possible that $a_1 = 0$ and $a_2 = +\infty$.

Definition 7.1. The system of N vector-solutions

$$\varphi_1(t),\ \varphi_2(t),\ \ldots,\ \varphi_N(t) \tag{7.2}$$

of the homogeneous system

$$\dot{y} = A(t)y \tag{7.3}$$

is called fundamental system of solutions, if those vectors are linearly independent.

It is known that for (7.3) there always exists fundamental system of solutions, then any its solution $y(t)$ can be presented in the form

$$y(t) = \sum_{j=1}^{N} C_i \varphi_j(t). \tag{7.4}$$

Proceeding from these vectors we form the matrix

$$\Phi(t) = \|\varphi_j^i(t)\|, \qquad i, j = 1, \ldots, N. \tag{7.5}$$

The function $W(t) = \det \Phi(t)$ is called Wronski determinant or Wronskian.

If (7.2) is a fundamental system of solutions, then $W(t) \neq 0$ for every value of t and matrix (7.5) corresponding to it is called fundamental matrix of solutions.

Let us rewrite any solution of (7.4) in the matrix form

$$y(t) = \Phi(t)C, \tag{7.4'}$$

where $C = (C_1, \ldots, C_N)$ is a constant vector.

7.1. Equation of an arbitrary order

Consider the following equation of an arbitrary order

$$P_N(\partial_t)x(t) \equiv x^{(N)} + a_1(t)x^{(N-1)} + \cdots + a_N(t)x = b(t), \tag{7.6}$$

where the coefficients $a_j(t)$ and the right-hand side $b(t)$ are continuous over (a_1, a_2). Let us introduce unknown functions

$$y_1 = x, \qquad y_2 = \dot{x}, \qquad \ldots, \qquad y_N = x^{(N-1)}.$$

The new functions satisfy the linear system of the form (7.1) with the right-hand side $\vec{b}(t) = (0, 0, \ldots, b(t))$ and with the matrix

$$A = \begin{Vmatrix} 0 & & & & \\ 0 & & I^{(N-1)} & & \\ \cdots & & & & \\ 0 & & & & \\ -a_N & -a_{N-1} & \cdots & -a_2 & -a_1 \end{Vmatrix}.$$

Thereby, equation (7.6) and system (7.1) are equivalent in the sense that to every solution $x = \psi(t)$ of the equation there corresponds the solution $y = \varphi(t) = (\psi(t), \dot{\psi}, \ldots, \psi_N^{(N-1)})$ of the system, and vice versa, to every solution $\varphi(t) = y = (\varphi_1, \varphi_2, \ldots, \varphi_N)$ of the system there corresponds the solution $x = \varphi_1(t)$ of the equation, and this correspondence is one-to-one.

In view of this dependence between the equation and the system, fundamental matrix of solutions of the obtained system can naturally be considered to be also a fundamental matrix of solutions of this equation of a higher order

$$\Phi(t) = \|x_j^{(k)}(t)\|, \qquad j = 1, \ldots, N, \quad k = 0, \ldots, N-1. \tag{7.7}$$

Since the system of solutions $x_j(t)$ is linearly independent, evidently $\det \Phi \neq 0$. The letter can be shown directly using the Liouville's formula

$$\det \Phi(t) = \exp\left[(t - t_0) \operatorname{tr} A\right] \det \Phi(t_0) = \det \Phi(t_0) \exp\left[-a_1(t)(t - t_0)\right].$$

Further let us study the inverse process, i. e. answer wither the first-order system is reducible to the equations. This question is more complex and system (7.3) is not always equivalent to one equation. The result depends on whether the matrix A is similar to the Frobenius form consisting of one cell P_1 or isn't.

And how can we find it out? The following facts are known (Godunov, 1986)

1. If eigenvalues of the matrix A are different, then $m = 1$, because in this case $r_j = N - 1$ for every j. At the same time, the transforming matrix T can be selected in the form $T = [b, Ab, \ldots, A^{N-1}]$, where b is any vector provided that there held the condition of the matrix T nondegeneracy.

2. The necessary condition for the Frobenius form to have several cells is the existence of multiple eigenvalues.

Let us construct a Gram's matrix

$$X = \|x_{ij}\|_{j,i=1}^{N}, \qquad x_{ij} = \int_{t'}^{t''} x_i(t)x_j(t)\,\mathrm{d}t, \quad [t', t''] \in [a, b].$$

Statement 7.1. Gram's matrix constructed on an arbitrary interval $[t', t'']$ from any fundamental system of solutions of equation (7.6) is defined strictly positively.

Let the coefficients $a_j = \text{const}$, then the characteristic polynomial can be factorized

$$P_N(\partial_t) = (\tau - \tau_1)^{k_1}(\tau - \tau_2)^{k_2} \ldots (\tau - \tau_S)^{k_s},$$
$$k_1 + k_2 \cdots + k_s = N, \qquad \tau_l \neq \tau_m, \quad l \neq m,$$

where τ_j are roots with the multiplicity k_j. Then the equation itself will be written in the form

$$P_N(\partial_t)x(t) \equiv (\partial_t - \tau_1)^{k_1} \ldots (\partial_t - \tau_s)^{k_s} x(t) = 0.$$

It can be stated that the set of function

$$\begin{cases} x_1^1(t) = \mathrm{e}^{\tau_1 t}, \quad x_2^1(t) = \dfrac{t}{1!}\,\mathrm{e}^{\tau_1 t}, \quad \ldots, \quad x_{k_1}^1(t) = \dfrac{t^{k_1-1}}{(k_1-1)!}\,\mathrm{e}^{\tau_1 t}; \\ x_1^2(t) = \mathrm{e}^{\tau_2 t}, \quad x_2^2(t) = \dfrac{t}{1!}\,\mathrm{e}^{\tau_2 t}, \quad \ldots, \quad x_{k_2}^2(t) = \dfrac{t^{k_2-1}}{(k_2-1)!}\,\mathrm{e}^{\tau_2 t}; \\ \cdots\cdots\cdots\cdots\cdots\cdots \\ x_1^s(t) = \mathrm{e}^{\tau_s t}, \quad x_2^s(t) = \dfrac{t}{1!}\,\mathrm{e}^{\tau_s t}, \quad \ldots, \quad x_{k_s}^s(t) = \dfrac{t^{k_s-1}}{(k_s-1)!}\,\mathrm{e}^{\tau_s t} \end{cases} \tag{7.8}$$

forms fundamental system of solutions.

7.2. The Green's matrix and construction methods

Let us start from the definition of the Green's matrix.

Definition 7.2. The continuously differentiable matrix $G_0(t, A) \equiv G_0(t)$ defined for every $t \neq 0$ is called the Green's matrix for system (7.1), if it

a) satisfies the system $\dot{G}_0(t,) \equiv AG_0(t)$ for $t \neq 0$;

b) has a discontinuity of the first kind at $t = 0$: $G_0(+0) - G_0(-0) = I$;

c) is limited for all $t \in \mathbb{R}$: $\|G_0(t)\| \leq C$.

Theorem 7.1. *For every A having no purely imaginary and zero eigenvalues there exists a unique Green's matrix $G_0(t)$, by means of which the solution of the inhomogeneous equation*

$$y(t) = \int_{\mathbb{R}} b(\tau)G_0(t - \tau)\,d\tau, \tag{7.9}$$

is given if $\|b(t)\| \leq \text{const}$ *for* $t \in \mathbb{R}$.

Let us show two methods for the construction of $G_0(t)$

1. Acceding to Schur's theorem for the matrix from Theorem 7.1 there exists a nonsingular matrix T such that

$$A = T^{-1}\begin{bmatrix} B & 0 \\ 0 & C \end{bmatrix} T,$$

where $\operatorname{Re}\lambda_B < 0$, $\operatorname{Re}\lambda_C > 0$.

Now let us find $G_0(t)$ by using the formula

$$G_0(t) = \begin{cases} T^{-1}\begin{bmatrix} e^{tB} & 0 \\ 0 & 0 \end{bmatrix} T, & t > 0, \\ T^{-1}\begin{bmatrix} 0 & 0 \\ 0 & -e^{tC} \end{bmatrix} T, & t < 0. \end{cases} \tag{7.10}$$

2. Polynomial representation has the form

$$G_0(t) = g(t, \tau_1)I + g(t, \tau_1, \tau_2)(A - \tau_1 I) + \cdots + g(t, \tau_1 \ldots, \tau_N) \prod_{j=1}^{N-1} (A - \tau_j I). \tag{7.11}$$

Here functions

$$g(t,\tau_1)=\begin{cases}\begin{cases}0, & t>0,\\ -\mathrm{e}^{\tau_1 t}, & t<0,\end{cases} & \operatorname{Re}\tau_1>0,\\[2ex] \begin{cases}\mathrm{e}^{\tau_1 t}, & t>0,\\ 0, & t<0,\end{cases} & \operatorname{Re}\tau_1<0,\end{cases}$$

$$g(t,\tau_1,\dots,\tau)=\begin{cases}\int_{-\infty}^{t} g(s,\tau_1,\dots,\tau_{k-1})\mathrm{e}^{\tau_k(t-s)}\,\mathrm{d}s, & \operatorname{Re}\tau_k<0,\\ -\int_t^{\infty} g(s,\tau_1,\dots,\tau_{k-1})\mathrm{e}^{\tau_k(t-s)}\,\mathrm{d}s, & \operatorname{Re}\tau_k>0,\end{cases} \tag{7.12}$$

or, which is the same

$$g(t,\tau_1,\dots,\tau_k)=\int_{-\infty}^{\infty} g(s,\tau_1,\dots,\tau_{k-1}),g(t-s,\tau_k)\,\mathrm{d}s.$$

Note two important properties of the above-mentioned functions:

a) if $\operatorname{Re}\tau_j<0$ for every $j=1,\dots,N$, then $G_0(t)=\begin{cases}\mathrm{e}^{tA}, \ t>0\\ 0, \ t<0\end{cases}$, i.e. it coincides with the matrix exponent;

b) the function $g(t,\tau_1)$ is piecewise continuous and has only one point of discontinuity $t=0$: $[g(0,\tau_1)]\equiv g(+0,\tau_1)-g(-0,\tau_1)=1$, all the remaining functions are continuous.

Chapter 1.

Construction of ill-posed problems for a system of the first order ODE

1.1. CORRECT PROBLEM STATEMENT FOR AN ARBITRARY SYSTEM OF ODE

Let us consider problem on the ray $t > 0$ (Solonnikov, 1965)

$$\begin{aligned} A(d)u(t) &= 0, \qquad t > 0, \\ B(d)u(t)\big|_{t=0} &= \varphi, \qquad |u| \to 0, \quad t \to \infty. \end{aligned} \tag{1.1.1}$$

Here A is a square matrix of order N, whose elements are the linear differential operators $\alpha_{sj}(d)$, of order arbitrarily dependent on the subscripts s and j. It is assumed that that the polynomial $A(\lambda) = \det A(\lambda)$ has no purely imaginary and zero roots, or, to be more precise, has k roots λ_s^- with a negative real part and m roots λ_s^+ with a positive ($N = k + m$). The matrix $B(d)$ is rectangular and let r be the number of its rows. Let us find the number r and determine the algebraic conditions, which the matrix B must satisfy for the unique solvability of problem (1.1.1) for every φ.

Let us represent the polynomial matrix $A(\lambda)$ as the product

$$A(\lambda) = U(\lambda)Q(\lambda)V(\lambda),$$

where U, Q, V are polynomial matrices, such that $\det U(\lambda) = \det V(\lambda) = 1$

and $Q(\lambda) = \operatorname{diag}(P_1(\lambda), \dots, P_N(\lambda))$. We have

$$A(\lambda) = \prod_{j=1}^{N} P_j(\lambda),$$

$P_j(\lambda)$ are invariant multinomials.

If we introduce a new function

$$W(t) = V(d)U(t), \qquad U(t) = V^{-1}(d)W(t),$$

then problem (1.1.1) is equivalent to the following problem

$$\begin{gathered} Q(d)W(t) = 0, \\ D(d)W\big|_{t=0} = \varphi, \qquad |W(t)| \to 0, \quad t \to \infty. \end{gathered} \tag{1.1.2}$$

Let $P_j(\lambda) = P_j^+(\lambda)P_j^-(\lambda)$, where P^+ and P^- are polynomials whose roots have positive and negative real parts, then (1.1.2) can be written in the form

$$\begin{gathered} P_j^-(d)W_j(t) = 0, \qquad j = 1, \dots, N, \\ D'(d)W(t)\big|_{t=0} = \varphi, \end{gathered} \tag{1.1.3}$$

elements of $D' = (d'_{qj})$ are residues of division of corresponding elements of the matrix D by $P_j^-(t)$.

Possibly, for some j the polynomials $P_j(t) = C_j \neq 0$ are constants (let them be such for $j = k' + 1, \dots, N$). Then they correspond to $W_j(t) = 0$ and $d_{qj} = 0$ (1.1.3). Let us construct a new vector from the remaining components

$$W' = (W_1, \dots, W_{k'}),$$

for which we have

$$\begin{gathered} P_j^-(d)W_j(t) = 0, \qquad j = 1, \dots, k'_j, \\ D''(d)W'(t)\big|_{t=0} = \varphi. \end{gathered} \tag{1.1.4}$$

Obviously, the matrix D'' is obtained from D' by truncating the last $(N - k')$ zero columns. Let us denote the degree of the multinomials $P_j^-(\lambda)$ by ω_j, then $k = \sum_{j=1}^{k'} \omega_j$ and $d'_{qj}(\lambda) = \sum_{s=1}^{\omega_j} C_{qj}^{(s)} \lambda^{s-1}$.

Rewrite the boundary condition from (1.1.4) in the form of system of linear algebraic equations (SLAE)

$$Cy = \varphi \tag{1.1.5}$$

where $y = (W_1(0), \dots, W_1^{(\omega_1-1)}(0), \dots, W_{k'}(0), \dots, W_{k'}^{(\omega_{k'}-1)}(0)$ is the desired vector, and

$$C = \begin{bmatrix} C_{11}^{(1)} \dots C_{11}^{(\omega_1)} \dots C_{1k'}^{(1)} \dots C_{1k'}^{(\omega_{k'})} \\ \dots\dots\dots\dots\dots\dots\dots \\ C_{r1}^{(1)} \dots C_{r1}^{(\omega_1)} \dots C_{rk'}^{(1)} \dots C_{rk'}^{(\omega_{k'})} \end{bmatrix}$$

is the $r \times k$ matrix.

Therefore for problem (1.1.4) to have a unique solution $W(t)$ for every φ it is necessary and sufficient that the system of linear equations (1.1.5) should have one solution y for every φ, and for the latter, in turn, it is necessary and sufficient that

$$r = k \qquad \text{and} \qquad \det C \neq 0. \tag{1.1.6}$$

It is proved (Solonnikov, 1965) that (1.1.6) is equivalent to Lopatinskii condition for problem (1.1.2) and therefore for the original problem (1.1.1).

Thus, the boundary matrix B should have the dimensionality $k \times N$. In addition, if there holds Lopatinskii condition (1.1.6), then (1.1.1) is uniquely solvable for every φ. If it is violated ($\det C = 0$), then it does not have solution (1.1.1) for the given φ, or has a whole family of solutions, because for the corresponding homogeneous problem ($\varphi = 0$) there exists a nontrivial solution.

THE FIRST BOUNDARY-VALUE PROBLEM

1.2. PROBLEM ON THE RAY $t > 0$

1.2.1. Lopatinskii condition. Reduction method

Consider a particular variant of problem (1.1.1). Find a bounded solution of an inhomogeneous normal first-order system

$$\begin{gathered} \dot{y} = Ay(t) + f(t), \qquad t > 0; \\ f(t) \text{ is continuous and } \|f(x)\| \leq F, \end{gathered} \tag{1.2.1}$$

for which the following equalities hold true

$$By(0) = \varphi. \tag{1.2.2}$$

Here the square matrix $A = (a_{ij})_{ij=1}^{N}$ has two kinds of eigenvalues

$$\begin{aligned} \operatorname{Re}\lambda_j &< -\delta, \qquad j = 1, \dots, k, \\ \operatorname{Re}\lambda_j &> \delta, \qquad j = k+1, \dots, N, \end{aligned} \tag{1.2.3}$$

$\delta > 0$, $k + m = N$, $1 \le k \le N$, $m \ge 0$; rectangular (constant) matrix $B = (b_{ij})$, $i = 1, \dots, k$, $j = 1, \dots, N$ of the rank k, satisfies one of the requirements: (M) (maximum condition) it has no proportional columns; (m)(minimum condition) there is no zero column in the matrix.

Further, we pursue two purposes: firstly, we give a simple analogue of the Lopatinskii condition (like (1.1.6)); secondly, we construct the solution of boundary value problem (1.2.1), (1.2.2), when it is ill-posed.

Let $\Phi(x)$ be fundamental matrix of solutions of the homogeneous system $\dot{y} = Ay$, and $\tilde{\Phi}(x)$ be its part corresponding to the first k eigenvalues ($\operatorname{Re}\lambda_j < -\delta$ and $\operatorname{rang}\tilde{\Phi}(0) = k$). The general solution of a homogeneous solution bounded for every $t > 0$ will be written as follows:

$$y(t) = \tilde{\Phi}(t)C, \tag{1.2.4}$$

where $C = (C_1, C_2, \dots, C_k)$ is an arbitrary constant vector.

By virtue of the boundary condition (1.2.2), we obtain the SLAE:

$$\mathcal{F}C = \varphi, \qquad \mathcal{F} = B\tilde{\Phi}(0). \tag{1.2.5}$$

Therefore, the differential problem is again reduced to the solution of a SLAE; hence we make obvious conclusions:

i) the criterion $\det\mathcal{F} \ne 0$ is the analogue of the Lopatinskii condition, because in this case system (1.2.5) has the unique solution C for every φ, consequently, problem (1.2.1), (1.2.2) is uniquely solvable.

ii) if $\det\mathcal{F} \ne 0$, but the quantity $|\det F|$ is close to zero, then the SLAE is ill-conditioned and thus Lopatinskii condition not always can provide continuous dependence of the solution on the input information, i. e. solution stability.

iii) if $\det\mathcal{F} = 0$, then the Lopatinskii condition is violated and therefore the problem is solvable not for all φ or solution is not unique.

We are interested in the latter case when the problem becomes ill-posed. First of all, a nontrivial solution of the homogeneous system $\mathcal{F}C = 0$ will be denoted by $\hat{C} = (\hat{C}_1, \dots, \hat{C}_K) \ne 0$.

The question is whether we can formulate an additional requirement which when fulfilled provides that $\hat{C} \equiv 0$ or it is unique. Then we first seek a normal or pseudo(normal) solution C^n of SLAE (1.2.4) determined uniquely, as is shown in Section 3 of the introductory chapter. After that using the found vector C^n we write the corresponding "solution" of the differential problem.

Thus, according to the theory of boundary-value problems (Bitsadze, 1976; Vladimirov, 1971; Godunov, 1971), we have:

$$y(t) = \tilde{\Phi}(t)C^n + \int_0^\infty G(t-s)f(s)\,\mathrm{d}s \tag{1.2.6}$$

is the solution of boundary value problem (1.2.1), (1.2.2);

$$y(t) = \tilde{\Phi}(t)\hat{C}$$

is the solution of the reduced homogeneous problem ($f = 0$, $\varphi = 0$), where $G(t)$ is the Green's matrix.

1.2.2. Density of ill-posed problems

Let us return to the main problem of this book: are there many ill-posed problems among the problems of the type (1.2.1), (1.2.2)? Earlier, we formulated the answer in the form of a hypothesis. Here we will turn it into a theorem.

Theorem 1.2.1. *For the fixed system* (1.2.1) *with the matrix A of class* (1.2.3) *there exists a boundary matrix B, which forms an ill-posed first boundary-value problem and, vice versa, there exists a system* (1.2.1) *with the matrix A of type* (1.2.3) *and for it the first boundary-value problem with the given matrix B will be ill-posed.*

Remark 1.2.1. Further the problem corresponding to the first part of the theorem will be called a direct problem and that corresponding to the second part will be named an inverse problem.

Proof. According to the reduction method, the proof will be reduced to the study of the SLAE:

$$\mathcal{F}C = \varphi, \qquad \mathcal{F} = (f_{ij}) = B\tilde{\Phi}(0) = \sum_{s=1}^{N} b_{is}v_{sj},$$

where $\tilde{\Phi}(0) = (v_{ij})$, $i = 1, \ldots, N$, $j = 1, \ldots, k$.

Thus, we would like to know when $\det \mathcal{F} = 0$. Let us separately consider the cases $k = 1$ and $2 \leq k < N$. If $k = 1$, then the matrices B and $\tilde{\Phi}(0)$ will become vector, which will require their orthogonality ($f_{11} = \sum_{s=1}^{N} b_{1s} v_{s1} = 0$). This is always possible, since one of them is an arbitrary (free) vector.

For the case $k \geq 2$, we can use one of the sufficient conditions of making the determinant vanish. That is, let the rows of the matrix $\mathcal{F}$ with the numbers q and n be proportional:

$$f_{qp} = \mu f_{np}, \qquad p = 1, \dots, k, \quad 1 \leq q \neq n \leq k,$$

and the proportionality coefficient μ can be any number. We introduce the following notation in those equalities

$$z_j = b_{qj} - \mu b_{nj}, \qquad j = 1, \dots, N \tag{1.2.7}$$

and obtain the matrix equation

$$\tilde{\Phi}^{\top}(0) z = 0 \qquad \Big(\sum_{j=1}^{N} v_{jp} z_j = 0, \quad p = 1, \dots, k \Big). \tag{1.2.8}$$

Thereby, the proof of the first part of the theorem is tantamount to the solution of system (1.2.8). Since $N > k$ and $\operatorname{rang} \tilde{\Phi}(0) = k$, it has a family of solutions

$$z_j = \sum_{s=1}^{m} C_{js} z_{k_s}, \qquad j = 1, \dots, k, \tag{1.2.9}$$

depending on $m \geq 1$ parameters (free components). Now, proceeding from the obtained values z_{n_j} on the basis of (1.2.7) we determine the elements b_{ij} ($i = q, n$; $j = 1, \dots, N$) of the matrix B and of the numerical parameter μ, all in all $2N + 1$ constants. Evidently, there will be free elements among those. It will be sufficient to specify them and other elements of the matrix B so that the condition (M) and (m) be satisfied. The first part of the theorem is proved.

In order to find the desired system of ODE with the matrix A or in order to prove the second part of the theorem let us assume that F has proportional columns with the numbers q and n. This yields the system

$$Bx = 0, \qquad x_j = v_{jq} - \mu v_{jn}, \quad j = 1, \dots, N. \tag{1.2.10}$$

This system also has a family of solutions

$$x_{nj} = \sum_{S=1}^{m} C'_{jS} x_{k_s}, \qquad j = 1, \dots, N. \tag{1.2.9$'$}$$

Then from the calculated values of x_j on the basis of (1.2.10) we determine the elements v_{js}, $j = 1, \dots, N$, $s = q, n$ of the matrix $\tilde{\Phi}(0)$ and the parameter μ. The remaining elements of the matrix $\Phi(0)$ are given arbitrarily, provided only that there holds the equality $\operatorname{rang} \tilde{\Phi}(0) = k$.

Finally, from the obtained $\tilde{\Phi}(0)$ and allowing for the criterion of other eigenvalues $\operatorname{Re} x_j > \delta$, $j = k+1, \dots, N$ we reconstruct the matrix A itself. Theorem is fully proved. □

Remark 1.2.2. As is seen from the above reasoning, the set of the matrices $\{A, B\}$, representing an ill-posed first boundary value problem is determined non-uniquely, or, to be more precise, there is a whole family of them.

One more proof. The both parts of the theorem may be proved on the basis of only one of the systems (1.2.8) or (1.2.10), i. e. proceeding from one criterion of the matrix degeneracy, i. e., the proportionality of the columns or rows with the numbers q and n. By way of example, let us take system (1.2.3) as a basis, then in order to prove the first part of the theorem, it is necessary to find kN of its elements b_{ij} using the given vector $x = (x_1, x_2, \dots, x_N)$, or, in other words, it is sufficient to find k vectors $b_s = (b_{ij})$, $j = 1, \dots, N$ orthogonal to this vector x. Therefore, it is sufficient to solve one vector equation with N unknowns under the fixed number $s = \bar{s}$:

$$x_1 b_{\bar{s}1} + x_2 b_{\bar{s}2} + \cdots + x_N b_{\bar{s}N} = 0, \qquad 1 \le \bar{s} \le k. \tag{1.2.11}$$

Remark 1.2.3. Note in particular the case when the procedure employed for the proof of the theorem is not completed, namely, when the number μ cannot be found for which the right-hand sides of all the equalities (1.2.7) and (1.2.9) or (1.2.10) and (1.2.9′) are comparable corresponding to the similar components of the vectors z and x.

Consequently, for such a pair A and B there holds the Lopatinskii condition, in other words, such a pair makes a uniquely solvable problem.

Remark 1.2.4. Previously, for the sake of convenience, simple (sufficient) indicators of the degeneracy of the matrix $\mathcal{F}$ were used. Actually, the degeneracy criterion to be used is $\det F = 0$ (especially for the small dimensionality of the matrices), since the sufficient conditions can lead to the loss of ill-posed problem.

Construction of direct and inverse problems

Consider the class of the problems including Godunov's examples (Godunov, 1983).

General case. Let us specify the class of the matrices A $(k = 2,\ m = 1)$ with simple eigenvalues λ_j and let fundamental matrix of solutions have the form

$$\Phi(x) = (1, \lambda_j, \lambda_j^2)^\top e^{\lambda_j x}, \qquad j = 1, 2, 3. \tag{$*$}$$

A) Direct problem. From the system

$$\tilde{\Phi}^\top(0)z = 0: \qquad z_1 + \lambda_1 z_2 + \lambda_1^2 z_3 = 0, \quad z_1 + \lambda_2 z_2 + \lambda_2^2 z_3 = 0$$

we find

$$z_1 = \lambda_1\lambda_2 z_3, \qquad z_2 = -(\lambda_1 + \lambda_2)z_3. \tag{1.2.9a}$$

On the other hand, according to (1.2.7) we have

$$z_1 = b_{11} - \mu b_{21}, \qquad z_2 = b_{12} - \mu b_{22}, \qquad z_3 = b_{13} - \mu b_{23}. \tag{1.2.7a}$$

We separately consider two statements about the roots λ_1 and λ_2.

a) Real roots: $\lambda_1 = -a$, $\lambda_2 = -b$, $a > 0$, $b > 0$, $a \neq b$. Then from two equations

$$b_{11} - \mu b_{21} = ab(b_{13} - \mu b_{23}), \qquad b_{12} - \mu b_{22} = (a + b)(b_{13} - \mu b_{23}),$$

obtained from the comparison of (1.2.7a) with (1.2.9a) we are to find seven quantities (number μ and elements b_{ij}). In particular, assuming $\mu = 0$, we construct the matrix

$$B = \begin{bmatrix} ab & a+b & 1 \\ b_{21} & b_{22} & b_{23} \end{bmatrix},$$

where b_{2j}, $j = 1, 2, 3$ are arbitrary real numbers.

b) Complex-conjugate roots: $\lambda_{1,2} = -a + ib$, $a > 0$, $b \in \mathbb{R}$. Now the sought quantities are found from the equalities

$$b_{11} - \mu b_{21} = (a^2 + b^2)(b_{\bar{1}3}\mu b_{23}), \qquad b_{12} - \mu b_{22} = 2a(b_{13} - \mu b_{23}),$$

obtained using the same method as in the case a). By assuming $\mu = 0$ we obtain a set of the matrices

$$B = \begin{bmatrix} a^2 + b^2 & 2a & 1 \\ b_{21} & b_{22} & b_{23} \end{bmatrix}$$

with arbitrary elements b_{2j}.

B) Inverse problem. From the given system $Bx = 0$ we find

$$x_1 = \frac{\Delta_1}{\Delta_2} x_3, \qquad x_2 = \frac{\Delta_2}{\Delta} x_3,$$

where conventional notation is introduced for the determinants:

$$\Delta = b_{11}b_{22} - b_{12}b_{21} \neq 0, \quad \Delta_1 = b_{12}b_{23} - b_{13}b_{22}, \quad \Delta_2 = b_{13}b_{21} - b_{11}b_{23}.$$

By the similar procedure we get the equalities

$$1 - \mu = \frac{\Delta_1}{\Delta}(\lambda^2 - \mu\lambda_2^2), \qquad \lambda_1 - \mu\lambda_2 = \frac{\Delta_2}{\Delta}(\lambda^2 - \mu\lambda_2^2).$$

From them we find the roots λ_1, λ_2 and the parameter μ.

When $\Delta_2^2 = \Delta\Delta_1$ we have $\mu = 0$, $\lambda_1 = \Delta/\Delta_2 < 0$, λ_2 is arbitrary.

If $\mu = (\lambda_1\Delta_1 - \Delta_2)/(\lambda_2\Delta_1 - \Delta_2) \neq 0$, from the first equality we obtain

$$ab\Delta_1 + (ab)\Delta_2 + \Delta = 0 \quad \text{or} \quad (a^2 + b^2)\Delta_1 + 2a\Delta_2 + \Delta = 0$$

depending on whether the roots λ_1, λ_2 are real or complex. After deriving a and b from them, we can find a matrix $\tilde{\Phi}(0)$ of the form $(*)$.

S. K. Godunov's examples (Godunov, 1983). The examples correspond to the complex-conjugate roots. Let the matrix

$$A = \begin{bmatrix} 0 & 1 & 0 \\ 0 & 0 & 1 \\ 1 & 0 & 0 \end{bmatrix}$$

be given for which $\lambda_{1,2} = (-1 \pm i\sqrt{3})/2$, $\lambda_3 = 1$. Besides

$$\tilde{\Phi}^\top(0) = \begin{bmatrix} 1 & \lambda_1 & \lambda_2 \\ 1 & \lambda_2 & \lambda_1 \end{bmatrix}^\top .$$

A′) Direct problem. We have (from calculation $z_1 = z_2 = z_3$)

$$b_{11} - \mu b_{21} = b_{12} - \mu b_{22} = b_{13} - \mu b_{23}.$$

In particular, assuming $\mu = 0$, we obtain a set of the matrices $B = \begin{pmatrix} 1 \\ b_{2j} \end{pmatrix}$ containing the matrix from Godunov's example $B^G = \begin{pmatrix} 1 & 1 & 1 \\ 1 & 0 & -1 \end{pmatrix}$, which defines the ill-posed problem. For the matrix $B_1 = \begin{pmatrix} 1 & -1 & 1 \\ 1 & 0 & -1 \end{pmatrix}$ the above-derived equalities assuming the form $1-\mu = -1 = 1+\mu$ are inconsistent for any μ (this matrix was also presented by Godunov).

Thus, the matrices A and B^G generate ill-posed problem since

$$\det \mathcal{F}^G = \det [B^G \tilde{\Phi}(0)] = (1+\lambda_1+\lambda_2)(\lambda_2-\lambda_1) = 0.$$

Actually, $\mathcal{F}^G$ has a zero row ($\mu = 0$).

For the pair A and B_1 there holds the Lopatinskii condition:

$$\begin{aligned}\det \mathcal{F}_1 &= \det [B_1 \tilde{\varphi}(0)] \\ &= (1-\lambda_1+\lambda_2)(1-\lambda_1) - (1-\lambda_2+\lambda_1)(1-\lambda_2) = -4\sqrt{3i} \neq 0.\end{aligned}$$

B′) Inverse problem. Find the matrix A in the selected class ($N = 3$, $k = 2$, λ_j are simple) such that for the system of ODE specified by it, the first boundary value problem with the boundary matrix B_1 is ill-posed.

Solution

Step 1. From system $B_1 x = 0$:

$$\begin{cases} x_1 - x_2 + x_3 = 0, \\ x_1 - x_3 = 0 \end{cases}$$

we obtain $x_1 = x_3$ and $x_2 = 2$.

Step 2. 1) The system of equations $1 = 1-\mu$; $2 = \lambda_2 - \mu\lambda_2$; $1 = \lambda_1^2 - \mu\lambda_2^2$ corresponding to the values $q = 1$, $n = 2$ is unsolvable (matrix $\mathcal{F}$ has no proportional rows).

2) Consider the equality $\det \mathcal{F} = 0$: $\lambda_1\lambda_2 - 2(\lambda_1+\lambda_2) + 1 = 0$, which is the degeneracy criterion of the matrix.

a) hence if we assume that the roots are real $\lambda_1 = -a < 0$, $\lambda_2 = -b < 0$, we have $a(b+2)+2b+1 = 0$, i. e., this yields the value $= -(2b+1)/(b+2) < 0$ of the opposite sign.

c) the case $\lambda_{1,2} = -a \pm ib$, where $a > 0$, $b \in \mathbb{R}$ is also impossible, because we obtain the equality $a^2+4a+b^2+1 = 0$, which has no desired value ($a > 0$). Thus, among the systems of ODE with fundamental matrix of solutions of the form (*) there is no solution of the inverse problem. Therefore, let us expand the set of ODE systems. To this end, we write the equalities from (1.2.10)

$$v_{11} - \mu v_{12} = 1, \qquad v_{21} - \mu v_{22} = 2, \qquad v_{31} - \mu v_{32} = 1,$$

corresponding to the general form of fundamental matrix of solutions. We select one of the families of solutions: let $\mu = 0$, $v_{11} = v_{12} = 1$, then $v_{21} = 2, v_{31} = 1$, other elements v_{22}, v_{32} are still free. We have only to complement this matrix $\tilde{\Phi}(0)$ to the complete (square) nondegenerate matrix

$$\Phi(0) = \begin{bmatrix} 1 & 1 & 1 \\ 2 & v_{22} & 1 \\ 1 & v_{32} & 1 \end{bmatrix} = V.$$

If we require $\det \Phi(0) = 1$, we find $v_{32} = 2$, and let the last arbitrary element $v_{22} = 3$.

Step 3. Reconstruction of the matrix A.

We specify the Jordan form $J = \operatorname{diag}(-1, -2, 2)$, then

$$A = VJV^{-1} = \begin{bmatrix} 1 & 1 & 1 \\ 2 & 3 & 1 \\ 1 & 2 & 1 \end{bmatrix} J \begin{bmatrix} 1 & 1 & -2 \\ -1 & 0 & 1 \\ 1 & -1 & 1 \end{bmatrix} = \begin{bmatrix} 3 & -3 & 2 \\ 6 & -4 & 0 \\ 5 & -3 & 0 \end{bmatrix}$$

Since eigenvalues are different, the system with the given matrix A can be reduced to one 3rd order equation. The transforming matrix T can be specified as follows: $= (b, Ab, Ab^2)$, where b is any vector, provided $\det T \neq 0$. For example

$$T = \begin{bmatrix} -2 & 3 & 7 \\ -3 & 0 & 18 \\ 0 & -1 & 15 \end{bmatrix} \qquad (\det T = 120),$$

$$T^{-1} = \frac{1}{120} \begin{bmatrix} 18 & -52 & 54 \\ 45 & -30 & 15 \\ 3 & -2 & 9 \end{bmatrix}, \qquad P = T^{-1}AT \begin{bmatrix} 0 & 0 & 4 \\ 1 & 0 & 4 \\ 0 & 1 & -1 \end{bmatrix}$$

consequently, $x''' + x'' - 4x' - 4x = 0$

Hence, according to the theory on the connection between the first-order system and the arbitrary equation, we can write the matrices

$$A_1 = \begin{bmatrix} 0 & 1 & 0 \\ 0 & 0 & 1 \\ 4 & 4 & -1 \end{bmatrix}, \qquad V_1 = \begin{bmatrix} 1 & 1 & 1 \\ -1 & -2 & 2 \\ 1 & 4 & 4 \end{bmatrix}.$$

Step 4. Verification of solutions

1) $P_3(x) = P_3^1(x) = \lambda^3 + \lambda^2 - 4\lambda - 4 = 0$ $(\lambda_1 = -1,\ \lambda_2 = -2,\ \lambda_3 = 2)$;

2)

$$\Phi(x) = \begin{bmatrix} e^{-x} & e^{-2x} & e^{2x} \\ 1 & 1 & 1 \\ 2 & 3 & 1 \\ 1 & 2 & 1 \end{bmatrix}, \qquad A\Phi(x) = \begin{bmatrix} e^{-x} & e^{-2x} & e^{2x} \\ -1 & -2 & 2 \\ -2 & -6 & 2 \\ -1 & -4 & 2 \end{bmatrix},$$

$$\Phi_1(x) = \begin{bmatrix} e^{-x} & e^{-2x} & e^{2x} \\ 1 & 1 & 1 \\ -1 & -2 & 2 \\ 1 & 4 & 4 \end{bmatrix}, \qquad A\Phi_1(x) = \dot{\Phi}_1(x) = \begin{bmatrix} e^{-x} & e^{-2x} & e^{2x} \\ -1 & -2 & 2 \\ 1 & 4 & 4 \\ -1 & -8 & 8 \end{bmatrix};$$

3)

$$B_1\tilde{\Phi}(0) = \begin{pmatrix} 1 & -1 & 1 \\ 1 & 0 & 1 \end{pmatrix} \begin{pmatrix} 1 & 1 \\ 2 & 3 \\ 1 & 2 \end{pmatrix} = \begin{pmatrix} 0 & 0 \\ 0 & -1 \end{pmatrix}, \quad B_1\tilde{\Phi}_1(0) = \begin{pmatrix} 3 & 7 \\ 0 & -3 \end{pmatrix}.$$

Conclusion:

i) As has been noted above, it would be better to use the degeneracy criterion of the matrix ($\det F = 0$).

ii) A particular form (*) of fundamental matrix of solutions can also lead to the loss of the solution of the inverse problem.

1.2.3. Solution of the problem with the matrices A and B^G

To the pair A and B^G there corresponds a special matrix

$$\mathcal{F} = \begin{bmatrix} 0 & 0 \\ (3+i\sqrt{3})/2 & (3-i\sqrt{3})/2 \end{bmatrix}. \tag{1.2.5a}$$

As is evident, if $\varphi_1 \neq 0$, then the system $\mathcal{F}C = \varphi$ is unsolvable, and if $\varphi_1 = 0$, then the family of solutions is of the form

$$C_1 = \frac{2\varphi_2 - (3-i\sqrt{3})C_2}{3+i\sqrt{3}}. \tag{1.2.12}$$

When C_1, C_2 are real, this equality breaks up into two independent equations

$$\begin{aligned} C_1 + C_2 &= 2\varphi_2/3, \\ C_1 &= C_2, \end{aligned}$$

i. e. $C_1 = C_2 = \varphi_2/3$.

To these constants there corresponds solution (1.2.6) of the differential problem

$$y(t) = \frac{e^{-t/2}}{3} \left\| \begin{matrix} 2\cos\theta \\ -\cos\theta - \sqrt{3}\sin\theta \\ \sqrt{3}\sin\theta - \cos\theta \end{matrix} \right\| \varphi_2 = \operatorname{Re}\tilde{\Phi}(t)\,\frac{\varphi_2}{3}, \tag{1.2.6a}$$

where $\theta = \sqrt{3}\,t/2$.

Remark. An additional requirement (mentioned in Subsection 1.1), which provides that $\hat{C} \equiv 0$ is solution of a homogeneous system, is that components C_j are real.

By the definition of pseudosolution we obtain (Tikhonov and Arsenin, 1986)

$$\begin{aligned} \|\mathcal{F}C - \varphi\|^2 &= \|\sqrt{3}\,(\sqrt{3}+i)C_1/2 + \sqrt{3}\,(\sqrt{3}-i)C_2/2 - \varphi_2\|^2 \\ &= |3(C_1 + C_2)/2 - \varphi_2 + i\sqrt{3}\,(C_1 - C_2)/2|^2 \\ &= (3(C_1 + C_2)/2 - \varphi_2)^2 + 3(C_1 - C_2)^2/4 \equiv f(C_1, C_2) \Rightarrow \min. \end{aligned}$$

Hence we find the system

$$\begin{cases} 2C_1 + C_2 = \varphi_2, \\ 2C_2 + C_1 = \varphi_2 \end{cases}$$

with the solution $C_1 = C_2 = \varphi_2/3$.

In this case $\min f(\varphi_2/3, \varphi_2/3) = 0$, which confirms the well-known fact: if there exists the exact (classical) solution, then pseudosolution coincides with it.

Further, we will be interested in the cases when $\mathcal{F}C = \varphi$ is unsolvable ($\varphi_1 \neq 0$) and when its solution is nonunique ($\varphi_1 = 0$, the sought C_1, C_2 are complex numbers).

When $\varphi_1 \neq 0$, we seek the pseudosolution:

$$\min_{C_j} J(C_1, C_2) = \min_{C_j} \|FC - \varphi\|^2$$

$$= \left\| \begin{matrix} -\varphi_1 \\ (3 + i\sqrt{3})C_1/2 + (3 - i\sqrt{3})C_1/2 - \varphi_2 \end{matrix} \right\|^2$$

The equalities connected with the extremum conditions $\partial_{C_1} J = 0$, $\partial_{C_2} J = 0$ form one equation $2C_1 + (1 - i\sqrt{3})C_2 = (1 - i/\sqrt{3})\varphi_2$. Hence

$$C_1 = (1 - i/\sqrt{3})\varphi_2/2 - (1 - i\sqrt{3})C_2/2. \qquad (1.2.12')$$

From the condition $\partial_{C_2} I = 0$ where

$$I(C_2) = (1 - i\sqrt{3})C_2^2/2 + 2i\varphi_2 C_2/\sqrt{3} + (1 - i\sqrt{3})\varphi_2^2/6,$$

we find the normal pseudosolution: $C_{2p} = (3 - i\sqrt{3})\varphi_2/6$ and in view of (1.2.12′) $C_{1p} = (3 + i\sqrt{3})\varphi_2/6$.

Let now $\varphi_1 = 0$, then assuming that the element $x^1 = 0$ from the definition of normal solution is fixed, we have (see (6.4))

$$\inf_{C_2} J(C_2) = \inf_{C_2} \|[2\varphi_2 - (3 - i\sqrt{3})C_2](3 + i\sqrt{3})^{-1}\|^2.$$

Whence using the standard method we obtain $C_2^n = (3 - i\sqrt{3})\varphi_2/6$ and by virtue of (1.2.12′) $C_1^n = (3 + i\sqrt{3})\varphi_2/6$. Therefore normal and pseudonormal solutions coincide.

Thus, in this variant formulas (1.2.6) yield the solutions

$$y_n(t) = \operatorname{Re} \tilde{\Phi}(t) C^n = \frac{e^{-t/2}}{\sqrt{3}} \left\| \begin{matrix} \sqrt{3}\cos\theta + \sin\theta \\ -2\sin\theta \\ -\sqrt{3}\cos\theta \end{matrix} \right\| \varphi_2 \quad \text{(the solution is unique)},$$

$$y(t) = \operatorname{Re} \tilde{\Phi}(t)\hat{C} = \frac{e^{-t/2}}{4} \left\| \begin{matrix} 2(2b - a\sqrt{3}\,)\sin\theta - 2\sqrt{3}\,b\cos\theta \\ (b + a\sqrt{3}\,)\sin\theta - 3(a - \sqrt{3}\,b)\cos\theta \\ (a\sqrt{3} - 5b)\sin\theta + (3a - \sqrt{3}b)\cos\theta \end{matrix} \right\|, \quad \theta = \frac{\sqrt{3}}{2}t.$$

Here, the solution of the homogeneous problem is determined with an accuracy of one arbitrary complex constant $\hat{C}_1 = -(1 - i\sqrt{3}\,)\hat{C}_2/2$, where $\hat{C}_2 = a + ib$.

1.2.4. Uniquely solvable variant

Theorem 1.2.2. *If A is a Gurvits' matrix ($k = N$), then for the first boundary value problem* (1.2.1), (1.2.2) *the Lopatinskii condition holds true.*

Proof. We argue by contradiction. Let the problem be ill-posed, whereas $\det B \neq 0$, then the homogeneous system (1.2.10) has only trivial solution $x \equiv 0$. From the determining components of the equalities $x_j = v_{jq} - \mu v_{jn} \equiv 0$ it follows that $\operatorname{rang} \tilde{\Phi}(0) = \operatorname{rang} \Phi(0) < N$, which is impossible. □

Let us explain the above. First, in this variant, the boundary value problem (1.2.1), (1.2.2) can readily be transformed into the Cauchy problem for system (1.2.1), which is always uniquely solvable. Second, when $k = N$ all the solutions of the system $\dot{y} = Ay$ are bounded.

1.3. PROBLEM ON THE INTERVAL

1.3.1. Criterion for the Lopatinskii condition violation

Let us consider the problem on the interval $[0, 1]$: find the solution of the inhomogeneous system

$$\dot{y} = Ay(t) + f(t), \qquad 0 < t < 1 \tag{1.2.1}$$

satisfying the inhomogeneous boundary conditions

$$Ly(0) = \varphi, \qquad Ry(1) = \psi. \tag{1.2.2'}$$

Note the difference of this problem from the previous problem on the ray $t > 0$. The boundary matrices $L = (l_{ij})$, $R = (r_{ij})$ are rectangular, with dimension $k \times N$ and $m \times N$ and the ranks k and m respectively, $\varphi = (\varphi_1, \ldots, \varphi_k)$, $\psi = (\psi_1, \ldots, \psi_m)$, $k + m = N$, $1 \leq k, m < N$ are specified constant vectors. Here, the number k does not necessarily coincide with the quantity of eigenvalues of the matrix A with the negative real part. We assume that if $\det \begin{Vmatrix} L \\ R \end{Vmatrix} \neq 0$, then for these matrices there holds the condition (M), and if $\det \begin{Vmatrix} L \\ R \end{Vmatrix} = 0$, then there holds the condition (m).

According to Godunov (1994), Kasymov (1999) the Lopatinskii condition is formulated in the form

$$\Delta = \det \begin{Vmatrix} L\varphi(0) \\ R\varphi(1) \end{Vmatrix} \neq 0. \tag{1.3.1}$$

Then violation of the Lopatinskii condition ($\Delta = 0$) will be the condition of ill-posedness. The analogue of Theorem 1.2.1 holds true.

Theorem 1.2.1′. *For the given system of differential equations* (1.2.1) *with the matrix of class* (1.2.3) *there can be found a pair of boundary matrices L and R, which constitute an ill-posed problem and, vice versa, there exist differential equations* (1.2.1) *with the matrix of type* (1.2.3), *for which problem* (1.2.1), (1.2.2′) *with the fixed pair of matrices L and R is ill-posed.*

Proof. Let us write out the determinant (1.3.1) for the case when the eigenvalues λ_j of the matrix A are simple and real

$$\Delta = \begin{vmatrix} \sum_{j=1}^N l_{1j} v_{j1} & \sum_{j=1}^N l_{1j} v_{jN} \\ \dots & \dots \\ \sum_{j=1}^N l_{kj} v_{j1} & \sum_{j=1}^N l_{kj} v_{jN} \\ e^{\lambda_1} \sum_{j=1}^N r_{1j} v_{j1} & e^{\lambda_N} \sum_{j=1}^N r_{1j} v_{jN} \\ \dots & \dots \\ e^{\lambda_1} \sum_{j=1}^N r_{mj} v_{j1} & e^{\lambda_N} \sum_{j=1}^N r_{mj} v_{jN} \end{vmatrix}.$$

Let this determinant have proportional columns with the numbers q and n:

$$\sum_j l_{sj} v_{jq} = \mu \sum_j l_{sj} v_{jn}, \qquad s = 1, \dots, k,$$
$$\sum_j r_{pj} v_{jq} = e^{\lambda_n - \lambda_q} \mu \sum_j r_{pj} v_{jn}, \qquad p = 1, \dots, m. \tag{1.3.2}$$

If we introduce two vectors x and y with the components

$$x_j = v_{jq} - \mu v_{jn}, \qquad y_j = v_{jq} - \mu e^{\lambda_n - \lambda_q} v_{jn}, \qquad j = 1, \dots, N, \tag{1.3.3}$$

then arrive at the matrix equations

$$Lx = 0 \qquad (k \text{ equations}),$$
$$Ry = 0 \qquad (m \text{ equations}). \tag{1.3.4}$$

Thereby, the proof of the first part of the theorem is tantamount to finding kN elements l_{ij} and mN elements r_{ij} from (1.3.4) using the given values of the components x_j and y_j (1.3.3) or, which is the same, to the determination of k vectors $l_s = (l_{sj})$ and m vectors $r_s = (r_{sj})$, orthogonal to the above-mentioned vectors x and y respectively. Actually, it is sufficient to solve two equations with N unknowns when the subscripts s and p are fixed:

$$s = \bar{s}: \qquad x_1 l_{\bar{s}1} + x_2 l_{\bar{s}2} + \cdots + x_N l_{\bar{s}N} = 0,$$
$$p = \bar{p}: \qquad y_1 r_{\bar{p}1} + y_2 r_{\bar{p}2} + \cdots + y_N r_{\bar{p}N} = 0.$$

Consequently, the matrix $\left\| \begin{matrix} L \\ R \end{matrix} \right\|$, representing an ill-posed problem, exists, and it is non-unique.

In order to prove the second part of the theorem, we solve the same system (1.3.4) but with respect to the vectors x and y under specified elements l_{sj} and r_{pj}. Since $N > K$ and $N > m$, equation (1.3.4) has a family of solutions with free (m or k) components. Thus, to the given class of boundary conditions (or, to be more precise, to the given pair of matrices L and R) there corresponds a set of ODE systems with the matrix A. □

Remark 1.3.1. It remains to reconstruct the matrix A from the given eigenvalues λ_n, λ_q and from eigenvectors v_{k^*}, $k = 1, \ldots, N$, determined from (1.3.3).

1.3.2. Method for solution

First consider some data from by Godunov (1994), Kasymov (1999).

Definition 1.3.1. Three matrices $G(t,s)$ $(N \times N)$, $G_L(t)$ $(N \times k)$, $G_R(t)$ $(N \times m)$ are called Green's matrices of boundary value problem (1.2.1), (1.2.2′), if these matrices satisfy the following conditions:

$$\partial_t G(t,s) = AG(t,s) \quad (t \neq s), \qquad LG(0,s) = 0, \qquad RG(1,s) = 0,$$
$$G(s+0,s) - G(s-0,s) = I_j^{(N)};$$
$$\partial_t G_L(t) = AG_L(t), \qquad LG_L(0) = I^{(k)}, \qquad RG_L(1) = 0;$$
$$\partial_t G_R(t) = AG_R(t), \qquad LG_R(0) = 0, \qquad RG_R(1) = I^{(m)}.$$

Following the paper by Kasymov (1999) the matrices $G_L(t), G_R(t)$ is called special matrices of boundary value problem (1.2.1), (1.2.2′) and denoted by a letter $U(t)$. In this paper one more matrix is introduced called the Cauchy matrix, which will be further used for constructing the matrix $G(t,s)$.

Definition 1.3.2. The matrix $K(t,s)$ of two variables is called the Cauchy matrix if

$$\partial_t K(t,s) = AK(t,s), \qquad K(s,s) = I^{(N)}.$$

By Kasymov (1999) the following theorem was proved:

Theorem. *If the Lopatinskii condition is hold, then solution of boundary value problem* (1.2.1), (1.2.2′) *is represented by the formula*

$$y(t) = G_L(t)\varphi + \int_0^1 f(s)G(t,s)\,\mathrm{d}s + G_R(t)\psi. \tag{1.3.5}$$

Kasymov (1999) explicitly constructed the Green's matrices entering (1.3.5). Obviously, when the Lopatinskii condition is violated the methods suggested there lose sense. Therefore, a new approach is necessary. The special matrix $U(t)$ is sought as

$$U(t) = \Phi(t)D, \tag{1.3.6}$$

where $D = (d_{ij})$ is the square constant matrix that should be defined.

The substitution of (1.2.8) into the boundary conditions from Definition 1.3.1 yields the following SLAE

$$\mathcal{F}D = I^{(N)}, \qquad \mathcal{F} = \left\| \begin{matrix} LU(0) \\ RU(1) \end{matrix} \right\|. \tag{1.3.7}$$

Since $\Delta = \det \mathcal{F} = 0$ the obtained matrix will be denoted by D_n. We have

$$U_n(t) = \varphi(t)D_n. \tag{1.3.6'}$$

In the obvious way by partitioning the matrix $U_n(t)$ into G_{Ln} and G_{Rn}, we write an analogue of the formula for the Green's matrix:

$$G_n(t,s) = \begin{cases} -G_{Rn}RK(1,s), & t < s, \\ -G_{Rn}K(1,s) + K(t,s), & t > s, \end{cases}$$

where $K(t,s) = \Phi(t)\Phi^{-1}(s)$.

Now the solution of the ill-posed variant of problem (1.2.1), (1.2.2′), will be the following

$$U_n(t) = G_{Ln}(t)\varphi + G_{Rn}(t)\psi + \int_0^1 G_n(t,s)f(s)\,ds. \tag{1.3.5'}$$

It follows from the above procedure that the vector $u_n(t)$ fully satisfies system (1.2.1), but do not satisfy boundary conditions (1.2.2′), because the considered problem is ill-posed.

1.3.3. Examples

1.3.3.1. Problem construction

For the following system and right-hand sides

$$A = \begin{bmatrix} -1 & 0 & 0 \\ 0 & -2 & 0 \\ 2 & 0 & 1 \end{bmatrix}, \qquad \lambda = \begin{pmatrix} -1 \\ -2 \\ 1 \end{pmatrix}, \qquad f = \begin{pmatrix} 0 \\ 0 \\ 1 \end{pmatrix},$$

$$\psi = (0), \qquad \varphi = \begin{pmatrix} 1 \\ 0 \end{pmatrix}, \tag{1.2.1'}$$

we will write fundamental matrix of solutions $\Phi(t)$ and the determinant Δ:

$$\Phi(t) = \begin{bmatrix} e^{-t} & 0 & 0 \\ 0 & e^{-2t} & 0 \\ e^{-t} & 0 & e^{t} \end{bmatrix}, \qquad \Delta = \begin{vmatrix} e_{11} - e_{13} & l_{12} & l_{13} \\ l_{21} - l_{23} & l_{22} & l_{23} \\ e^{-1}(r_{11} - r_{13}) & e^{-2} r_{12} & e r_{13} \end{vmatrix}.$$

Direct problem. Let $q = 1$, $n = 2$, $\mu = 1$, then $l_{11} - l_{13} = l_{12}$, $l_{21} - l_{23} = l_{22}$, $l(r_{11} - r_{13}) = r_{12}$. These equalities hold true, for example, for a pair of the matrices $L = \begin{bmatrix} 2\,1\,1 \\ 1\,0\,1 \end{bmatrix}$ and $R = (1, 0, 1)$, satisfying the condition (m).

Inverse problem. It is easily verified that the boundary matrices $L_1 = \begin{bmatrix} 1\,1\,0 \\ 1\,1\,1 \end{bmatrix}$, $R_1 = (e, e^2, e)$ (for the system A) represent a well-posed problem ($\Delta \neq 0$). Let us find such a matrix A_1, that the triple A_1, L_1, R_1 form an ill-posed problem. Let its Jordan form be such that $J = \operatorname{diag}(-1,-2,1)$ and let Δ_1 have proportional rows $q = 1$, $n = 3$ with $\mu = 1$. We have

$$v_{21} = ev_{21} + v_{31}, \quad v_{12} = (v_{12} + v_{32})\mathrm{e}^{-1}, \quad v_{13} + v_{23} = \mathrm{e}^2(v_{13} + ev_{23} + v_{33}).$$

Hence by distinguishing the transformation matrix

$$V = \begin{bmatrix} 1 & 0 & \mathrm{e}^2 \\ 0 & 1 & 0 \\ 0 & 0 & 1-\mathrm{e}^2 \end{bmatrix},$$

we arrive at the desired matrix

$$A_1 = \begin{bmatrix} -1 & 0 & 2\mathrm{e}^2/(1-\mathrm{e}^2) \\ 0 & -2 & 0 \\ 0 & 0 & 1 \end{bmatrix},$$

which yields the solution of the inverse problem.

1.3.3.2. Solution of the problem A, L, R

Let us find the special matrix $U_n(t)$ (1.3.6′). For this purpose, we write out system (1.3.7)

$$\begin{gathered} d_{11} + d_{21} + d_{31} = 1, \qquad d_{12} + d_{22} + d_{32} = 0, \qquad d_{13} + d_{23} + d_{33} = 0, \\ d_{31} = 0, \qquad d_{32} = 1, \qquad d_{33} = 0, \qquad ed_{31} = ed_{32} = 0, \qquad ed_{33} = 1. \end{gathered}$$

It is obvious that this system is inconsistent. Therefore, we obtain its pseudosolution from the variational problem $J(d) = \|\mathcal{F}D - I\|^2 \to \inf_{d_{ij}}$.

The well-known extremum condition $\partial_{d_{ij}} J = 0$ generates the SLAE:

$$\begin{gathered} d_{11} + d_{21} + d_{31} = 1, \qquad d_{12} + d_{22} + d_{32} = 0, \qquad d_{13} + d_{23} + d_{33} = 0, \\ d_{11} + d_{21} + (2+\mathrm{e}^2)d_{31} = 1, \qquad d_{12} + d_{22} + (2+\mathrm{e}^2)d_{32} = 1, \\ d_{13} + d_{23} + (2+\mathrm{e}^2)d_{33} = e. \end{gathered}$$

Hence, we find a family of pseudosolutions

$$D^n = \begin{bmatrix} d_{11} & d_{12} & d_{13} \\ 1-d_{11} & -1/a & -e/a - d_{13} \\ 0 & 1/a & e/a \end{bmatrix},$$

here and below $a = 1 + \mathrm{e}^2$.

Define the normal pseudosolution $I(D) = \|D^n\|^2 \to \inf$. From the equality $\partial_{d_{ij}} I = 0$ we find their arbitrary values

$$d_{11} = 1/2, \qquad d_{13} = -e/2a, \qquad d_{12} = 0$$

which yields the special matrix

$$U_n(t) = G_{Ln}(t) + G_{Rn}(t) = \frac{\mathrm{e}^{-t}}{2a}\begin{bmatrix} a-1 \\ a\mathrm{e}^{-t} - \mathrm{e}^{-t} \\ -a+1+2\mathrm{e}^{2t} \end{bmatrix} + \frac{\mathrm{e}^{1-t}}{2a}\begin{bmatrix} -1 \\ -\mathrm{e}^{-t} \\ 1+2\mathrm{e}^{2t} \end{bmatrix}.$$

Calculate the Green's matrix

$$G_n(t,s) = \begin{cases} G_1(t,s), & t < s, \\ G_1(t,s) + K(t,s), & t > s, \end{cases}$$

where

$$G_1(t,s) = \frac{\mathrm{e}^{2-s-t}}{2a}\begin{bmatrix} 1 & 0 & 1 \\ \mathrm{e}^{-t} & 0 & \mathrm{e}^{-t} \\ -1-2\mathrm{e}^{2(t-s)} & 0 & -1-2\mathrm{e}^{2t} \end{bmatrix},$$

$$K(t,s) = \mathrm{e}^{s-t}\begin{bmatrix} 1 & 0 & 0 \\ 0 & \mathrm{e}^{s-t} & 0 \\ \mathrm{e}^{2(t-s)} - 1 & 0 & \mathrm{e}^{2(t-s)} \end{bmatrix}.$$

Finally, we find the vector

$$U_n(t) = G_{Ln}(t)\varphi + \int_0^t [G_1(t,s) + K(t,s)]f(s)\,\mathrm{d}s + \int_t^1 G_1(t,s)f(s)\,\mathrm{d}s$$
$$= \frac{1}{2}\begin{bmatrix} \mathrm{e}^{-t} + (\mathrm{e}^{2-t} - \mathrm{e}^{1-t})/a \\ \mathrm{e}^{-2t} + (\mathrm{e}^{2-2t} - \mathrm{e}^{1-2t})/a \\ \mathrm{e}^{-t} + 2\mathrm{e}^{t} - 2 + (\mathrm{e}^{1-t} - \mathrm{e}^{2-t} + 2\mathrm{e}^{1+t} - 2\mathrm{e}^{2+t})/a \end{bmatrix},$$

representing the solution of the ill-posed problem.

Actually, it satisfies this heterogeneous system of ODE (1.2.1′) and the boundary conditions

$$Lu_n(0) = \begin{bmatrix} 1 \\ (e - \mathrm{e}^2)/a \end{bmatrix}, \qquad Ru_n(1) = (l-1)/a$$

with different right-hand sides.

GENERAL BOUNDARY VALUE PROBLEM

1.4. PROBLEM STATEMENT. ANALYSIS OF BOUNDARY CONDITIONS

The boundary condition are given by the operator (differential) matrix

$$B(\partial) = \{b_{ij}\partial_x^{\alpha_{ij}}\}, \qquad i = 1,\dots,k, \quad j = 1,\dots,N. \tag{1.2.2''}$$

From nonnegative integer indices α_{ij} we form a new matrix $\alpha = (\alpha_{ij})$. Now problem (1.2.1), (1.2.2″) is characterized by three numerical matrices, i. e., $[A, B = B(1), \alpha]$.

1.4.1. Different and real eigenvalues of the matrix A

Notation. For definiteness and simplicity, let us assume that all eigenvalues of the matrix A are simple and real. Then the fundamental matrix of solution of system (1.1.1) $\dot{y} = Ay$ can be written in the form

$$\Phi(x) = \{v_{ij}l^{\lambda_j x}\}, \qquad V = \Phi(0) = \{v_{ij}\}, \quad \det V \neq 0. \tag{1.2.10n}$$

Further the following notation will be used: $B(\lambda)$ is a scalar matrix; $V(\lambda^{\alpha_k})$ is a matrix with the parameter α_k; $V(\lambda^{\alpha_k}) = \{v_{ij}\lambda_j^{\alpha_{kj}}\}$; $V_s = (v_{*s}) = (v_{1S},\dots,v_{Ns})$ is s-th column vector of V or s-th row of $V^\top$; $B_s = (b_{s*}) = (b_{s1},\dots,b_{sN})$ is the s-th row vector of B; $B^{=}_{s,\lambda_p}(b_{s*}\lambda) = (b_{s1}\lambda_p^{a_{S1}},\dots,b_{sN}\lambda_p^{a_{SN}})$ is the s-th row vector of $B(\lambda)$; $\lambda_* = (\lambda_1,\dots,\lambda_N)$ is the vector of the eigenvalues; $\alpha_s = (\alpha_{s*}) = (\alpha_{s1},\dots,\alpha_{sN})$ is s-th row vector of (α).

Let us consider the matrix

$$\mathcal{F} = B(\partial)\tilde{\Phi}(x)\big|_{x=0} = \Big\{\sum_{s=1}^{N} b_{is}\lambda_k^{\alpha_{is}}v_{sk}\Big\}_{i,k=1}^{N} = (f_{ik}).$$

Let us show its column and row notations

1) $F = \{B(\lambda_1)V_1,\dots,B(\lambda_q)V_q,\dots,B(\lambda_N)V_N\}$. As is clear, every q-th column of the matrix F is a product of the matrix $B(\lambda_q)$ by the vector column V_q.

$$2)\ F = \begin{Bmatrix} (B_{1,\lambda_s} \cdot V_s) \\ \cdots \\ (B_{q,\lambda_s} \cdot V_s) \\ \cdots \\ (B_{N,\lambda_s} \cdot V_s) \end{Bmatrix}_{s=1,\dots,N} \quad \text{or } 2')\ F = \begin{Bmatrix} V^\top(\lambda^{a_1})B_1 \\ \cdots \\ V^\top(\lambda^{a_q})B_q \\ \cdots \\ V^\top(\lambda^{a_N})B_N \end{Bmatrix}, \text{ where}$$

$V^\top(\lambda^{\alpha_q}) = \{V_s \lambda_s^{\alpha_q}\}_{s=1,\dots,N}$.

Every q-th row consists of scalar products of row vectors B_{q,λ_s} and column vectors V_s, $s = 1, \dots, N$, or, to be more precise, the s-th element of the fixed q-th row, i. e., element f_{qs}, is calculated as a scalar product of the column vector (V_{*s}) by the row vector B_{q,λ_k}, $f_{qs} = (V_{*s,b_{q*\lambda_k}})$ or the q-th row is the product of the transposed matrix $V^\top(\lambda^{\alpha_q})$ by the row vectors B_q.

1.4.2. Equality $\det \mathcal{F} = 0$. Possibility of construction of an ill-posed problem

We analyze the dependence of the criterion of degeneracy of $\mathcal{F}$ on the matrices B, V, α and the vector λ_*. Let us number eigenvalues of the matrix A the ascending order of their modules $|\lambda_1| < |\lambda_2| < \dots < |\lambda_N|$.

Let us use the sufficient criteria. Two cases cases are possible:

— the rows (columns) with the numbers q and n $(n > q)$ are proportional;

— there exists a "zero" column ("zero"-row).

Let us study these cases.

a) Let the column with the number q be "zero", then $\det B(\lambda_q) = 0$ and to the root λ_q there corresponds the vector V_{*q}, determined from the system $V(\lambda_q)V_q = 0$. Here the boundary matrix $B(\partial)$ is specified and then from the known values λ_q and V_q the Gurvits matrix A is reconstructed. In this case, we deal with the inverse problem.

b) If $\det V^\top(\lambda^{\alpha_q}) = 0$ and B_q is the vector corresponding to the "exponent vector" α_q, then the q-th row of $\mathcal{F}$ is "zero". In this case the direct problem is studied.

c) Let now the columns with the numbers q and n be proportional: $B(\lambda_q)V_q = \mu B(\lambda_n)V_n$, $\mu \neq 0$.

Formally, we return to the previous (known) form of notation $Bx = 0$, but here the values $x_s = \lambda_q^{\alpha_{is}} v_{sq} - \mu \lambda_n^{\alpha_{is}} v_{sn}$, $s = 1, \dots, N$ are quite different. Hence, since $\det B(1) \neq 0$ there follows the identical solution $x \equiv 0$. This implies that there holds the equality

$$v_{sq} = \mu b^{\alpha_{is}} v_{sn}, \tag{1.4.2}$$

where $b = \lambda_n/\lambda_q > 1$, $s, i = 1, \dots, N$.

If it is taken into account that for the fixed q, n and s the index i assumes any value from 1 to N, equalities (1.4.2) hold true only when, in the first place, the components of the vector $\alpha_s = (\alpha_{*s})$ assume the same values $\alpha_{is} = \beta_s$ for every $i \geq 1$, and in the second place, some (fixed) one-dimensional components of the vectors V_q and V_n are equal to zero ($v_{\bar{k}q} = v_{\bar{k}n} = 0$). In this case, the respective vectors $\alpha_{\bar{k}}$ can be arbitrary.

Consequently, if at least two constants β_s, $s = 1, \dots, N$ fail to coincide, the vectors V_q and V_n may be selected independent of each other. As a result, there holds

Lemma 1.4.1. *A class of direct ill-posed problems can be constructed with any nonsingular matrix* $B(1)$.

Proof. For $s = s_1$ and assuming $\beta_{s_1} = \alpha_{1s_1} = \dots = \alpha_{Ns_1} = 0$ we find $\mu_1 = v_{s_1q}/v_{s_1n} \neq 0$, otherwise $v_{s_1q} = v_{s_1n}$. For $s = s_2 \neq s_1$, let $\alpha_{1s_2} = \dots = \alpha_{Ns_2} = \beta_{s_2} \geq 1$. Then we have $v_{s_2q} = \mu_1 b^\beta v_{s_2n} \neq 0$, $b^\beta > 1$.

If we assume all the components $v_{kq} \equiv v_{kn} \equiv 0$ when $k \neq s_1$ and $k \neq S_2$. The missing columns V_s ($s \neq s_1$, $s \neq s_2$) of the matrix V are arbitrary. □

d) Now we study the case of proportional rows. We have

$$V^\top z = 0, \tag{1.4.3.I}$$

where $z_j = b_{qj}\lambda_s^{\alpha_{qj}} - \mu b_{nj}\lambda_s^{\alpha_{nj}}$, $j, s = 1, \dots, N$. Following the above-formulated principle we obtain the relation

$$b_{qj} = \mu b_{nj}\lambda_s^{\alpha_{nj}-\alpha_{qj}}, \qquad \mu \neq 0,$$

for every j at all $s \geq 1$.

Hence we arrive at the equalities

$$\lambda_1^{\beta_j} = \lambda_2^{\beta_j} = \dots = \lambda_N^{\beta_j}, \qquad \beta_j = \alpha_{nj} - \alpha_{qj}. \tag{1.4.2'}$$

They hold true only if $\beta_j = 0$. Consequently, $B_q = \mu B_n$ and this contradicts to the nondegeneracy of the matrix B.

Lemma 1.4.2. *For every matrix V the result is negative, in other words, the direct problem does not always exist.*

Conclusion. System (1.4.3) should be solved with respect to the vectors B_q and B_n but depending on the matrices V and α, i.e. by varying the elements v_{ij} and α_{ij} in their allowed ranges.

1.4.3. One of the eigenvalues is multiple

Let there be one eigenvalue of the matrix A with multiplicity p and let the Jordan form be as follows

$$J = \begin{bmatrix} \lambda_1 & \dots & 0 & 0 & 0 & \dots & 0 \\ \dots & \dots & \dots & \dots & \dots & \dots & \dots \\ 0 & \dots & \lambda_{n-1} & 0 & 0 & \dots & 0 \\ 0 & \dots & 1 & \lambda_n & 0 & \dots & 0 \\ 0 & \dots & 0 & 1 & \lambda_n & \dots & 0 \\ \dots & \dots & \dots & \dots & \dots & \dots & \dots \\ 0 & \dots & 0 & 0 & 0 & 1 & \lambda_n \end{bmatrix}.$$

Then we write fundamental matrix of solutions in the form

$$\Phi(x) = \Big\{ v_{sk} e^{\lambda_k x}, \ k = 1, \dots, n;$$
$$e^{\lambda_n x} \sum_{m=0}^{k} \frac{x^m}{m!} V_{n+k-m}, \ k = 1, \dots, p-1 \Big\}, \qquad (1.2.10k)$$

where (v_{si}), $i = 1, \dots, n$ are eigenvectors of the matrix A corresponding to simple eigenvalues λ_j, $j = 1, \dots, n-1$ and to the multiple eigenvalues λ_n, v_{si}, $i = n+1, \dots, N$ are adjoint vectors corresponding to p multiple eigenvalue λ_n.

Let us calculate the matrix of the boundary SLAE

$$\mathcal{F} = \sum_{j=1}^{N} b_{sj} d_x^{\alpha_{sj}} \Big\{ v_{jk} e^{\lambda_k x} \ (k = 1, \dots, n);$$
$$e^{\lambda_n x} \sum_{m=0}^{K} \frac{x^m}{m!} V_{n+k-m} \ (k = 1, \dots, p-1) \Big\} \Big|_{x=0} =$$

$$= \sum_{j=1}^{N} b_{sj} \Big\{ v_{jk} \lambda_k^{\alpha_{sj}} \ (k = 1, \dots, n) \Big[d^{\alpha_{sj}}$$

$$\sum_{m=0}^{K} \frac{x^m}{m!} V_{n+k-m} e^{\lambda_n x} \ (k = 1, \dots, p-1) \Big] \Big|_{x=0} \Big\}, \qquad s = 1, \dots, N.$$

In the further study the procedure is applied similar to the previous subsection.

Thus we have considered variants that do not defy a simple analysis and confirmed that for the Gurvits matrix A the statement of Theorem 1.2.1 remains valid if a boundary problem with the differential matrix $B(\partial)$ (1.2.2″) is considered. Namely, there holds

Theorem 1.4.1. *For the given (fixed) system* (1.2.1) *with the Gurvits matrix A there exists the operator boundary matrix $B(\partial_x)$, which forms an ill-posed boundary-value problem, and vice versa, there can be found a system of type* (1.2.1) *with the Gurvits matrix A and for it the boundary-value problem with this matrix $B(\partial_x)$ will be ill-posed.*

Remark 1.4.1. Difficulties arising in connection with the multiplicity of eigenvalues, may be overcome using two powerful instruments, i. e. the method of mathematical induction and passage to continuity limit. Therefore it is actually sufficient to confine the consideration to the case of simple roots.

1.4.4. Examples of sufficient conditions

Consider a general boundary-value problem for $k = 3$ $(k = N)$.

a) Let the first column be zero. Let us find the matrix A among the matrices similar to the Jordan form $J = \operatorname{diag}(-1, -2, -3)$. Take the matrix

$$B(\lambda) = \begin{bmatrix} \lambda & b_{12} & b_{13} \\ b_{21} & b_{22} & \lambda \\ b_{31} & \lambda & \lambda \end{bmatrix},$$

it implies that

$$\alpha = \begin{bmatrix} \alpha_1 \\ \alpha_2 \\ \alpha_3 \end{bmatrix} = \begin{pmatrix} 1 & 0 & 1 \\ 0 & 0 & 1 \\ 0 & 1 & 1 \end{pmatrix}.$$

Arbitrary elements b_{ij} we determine from the conditions $\det B(1) = 1$ and $\det B(-1) = 0$. These two equations contain five unknowns, therefore there is a set of solutions. We fix one of those ($b_{12} = -1$, $b_{13} = 3$, $b_{21} = 7/16$, $b_{22} = 2$, $b_{31} = 1/4$), calculate the eigenvector $V_1 = (16, -1, 5)$ and write accompanying matrices

$$V = \begin{bmatrix} 16 & 5/4 & 1 \\ -1 & 1 & 1 \\ 5 & -1 & -1 \end{bmatrix}, \qquad V^{-1} = \begin{bmatrix} 0 & 1/4 & 1/4 \\ 4 & -21 & -17 \\ -4 & 89/4 & 69/4 \end{bmatrix}.$$

Here free elements v_{ij} are fitted in such a way that $\det V = 1$. After the matrix A has been reconstructed, let us verify the ill-posedness of the constructed problem

$$A = V^{-1}JV = \begin{bmatrix} -13/4 & 1/4 & 1/4 \\ 149 & -14 & -13 \\ -601/4 & -49/4 & 45/4 \end{bmatrix},$$

$$\mathcal{F} = B(\partial)\Phi(x)\big|_{x=0} = \begin{pmatrix} \partial & -1 & 3 \\ 7/16 & 2 & \partial \\ 1/4 & \partial & \partial \end{pmatrix} \begin{bmatrix} \mathrm{e}^{-x} & \mathrm{e}^{-2x} & \mathrm{e}^{-3x} \\ 16 & 5/4 & 1 \\ -1 & 1 & 1 \\ 5 & -1 & -1 \end{bmatrix}$$

$$= \begin{pmatrix} 0 & x & x \\ 0 & x & x \\ 0 & x & x \end{pmatrix}.$$

b) Let now the first row be "zero". Let λ_* and α retain their previous values. The elements of the matrix B are sought proceeding from the equalities $\det V^\top(\lambda^{\alpha_1}) = 0$ and $V^\top(\lambda^{\alpha_1})B_1 = 0$. Let us write out the matrix $V^\top(\lambda^{\alpha_1})$ and specify its arbitrary elements

$$V^\top(\lambda^{\alpha_1}) = \begin{bmatrix} v_{11}\lambda_1 \\ v_{12}\lambda_2 v_{ij} \\ v_{13}\lambda_3 \end{bmatrix} = \begin{bmatrix} -v_{11} \\ -2v_{12}v_{ij} \\ -3v_{13} \end{bmatrix} = \begin{bmatrix} -16 & -1 & -8/53 \\ -5/2 & 1 & 1/2 \\ -3 & 1 & 27/53 \end{bmatrix}$$

$$(\det V = 1/4).$$

We solve the SLAE

$$\begin{bmatrix} -16 & -1 & -8/53 \\ -5/2 & 1 & 1/2 \\ -3 & 1 & 27/53 \end{bmatrix} \begin{bmatrix} b_{11} \\ b_{12} \\ b_{13} \end{bmatrix} = 0 \Rightarrow B = \begin{bmatrix} 1/53 & -24/53 & 1 \\ b_{21} & b_{22} & b_{23} \\ b_{31} & b_{32} & b_{33} \end{bmatrix}$$

$$\xrightarrow{\det B=1} B(\lambda) = \begin{bmatrix} \lambda/53 & -24/53 & 1 \\ 2 & 0 & 53\lambda \\ 1 & \lambda & 53\lambda/2 \end{bmatrix}.$$

The verification of the ill-posedness is

$$\mathcal{F} = \begin{pmatrix} \partial/53 & -24/53 & 1 \\ 2 & 0 & 53\partial \\ 1 & \partial & (53/2)\partial \end{pmatrix} \begin{bmatrix} e^{-x} & e^{-2x} & e^{-3x} \\ 16 & 5/4 & 1 \\ -1 & 1 & 1 \\ -8/53 & 1/2 & 27/53 \end{bmatrix} \Bigg|_{x=0}$$

$$= \begin{pmatrix} 0 & 0 & 0 \\ & & \\ & & \end{pmatrix},$$

at the same time

$$A = V^{-1}JV = \frac{1}{53} \begin{bmatrix} -66 & -48 & -25 \\ 672 & 1791 & 1912 \\ 632 & -1921 & -2043 \end{bmatrix}.$$

c) Let the first and the second columns ($q = 1$, $n = 2$) be proportional. Assuming $\alpha_{11} = \alpha_{21} = \alpha_{31} = 0$ ($s = 1$), we obtain $\mu = v_{11}/v_{12} = 1/2$ ($v_{11} = 1$, $v_{12} = 2$). Next, let $\alpha_{12} = \alpha_{22} = \alpha_{32} = 2$ ($\beta_2 = 2$), then $v_{21} = 2$, $v_{22} = 1$ ($b = 2$).

Now we can take $v_{31} = v_{32} = 0$. Thus,

$$V = \begin{bmatrix} 1 & 2 & v_{13} \\ 2 & 1 & v_{23} \\ 0 & 0 & -1/3 \end{bmatrix}, \qquad \det V = 1.$$

We have for an arbitrary matrix B ($\det B \neq 0$)

$$\mathcal{F} = \begin{bmatrix} b_{11} & b_{12}\partial^2 & b_{13} \\ b_{21} & b_{22}\partial^2 & b_{23} \\ b_{31} & b_{32}\partial^2 & b_{33} \end{bmatrix} \left(\begin{matrix} e^{-x} & e^{-2x} & e^{-3x} \\ 1 & 2 & v_{13} \\ 2 & 1 & v_{23} \\ 0 & 0 & -1/3 \end{matrix} \right) \Bigg|_{x=0}$$

$$= \begin{bmatrix} b_{11}+2b_{12} & 2(b_{11}+2b_{12}) & b_{11}v_{13}+9b_{12}v_{23}-b_{13}/3 \\ b_{21}+2b_{22} & 2(b_{21}+2b_{22}) & * \\ b_{31}+2b_{32} & 2(b_{31}+2b_{32}) & * \end{bmatrix}$$

1.5. LINEARIZATION ALGORITHM

So far, in particular cases we have solved SLAE with functional coefficients nonlinearly depending on elements of the vector λ and of the matrix α. Below, the algorithm for the linearization of these systems is given. Let us start from analyzing the variant d).

Thus, let the matrix $\mathcal{F}$ have proportional rows with the numbers q and n.

Step 1. Taking the set of eigenvalues $\{\lambda_*\}$, we specify the Jordan form of the matrix $A \div J_A = \operatorname{diag}(\lambda_1, \lambda_2, \ldots, \lambda_N)$, where $\lambda_j < 0$, $\lambda_j = 1, \ldots, N$, thereby we determine the class of the ODE system.

Step 2. By specifying the set of "exponents" (matrix α), we determine the class of boundary value problems.

This yields the homogeneous bilinear system of equations with parameter μ

$$(C_s^{(q)} B_q - \mu C_s^{(n)} B_n, V_s) = 0, \qquad s = 1, \ldots, N \tag{1.4.3$'$.I}$$

with respect to the pair (B_q, B_n) or the set of vectors V_s.

The components of the vectors, i. e. rows $C_s^{(q)} = (\lambda_s^{\alpha_q})$ and $C_s^{(n)} = (\lambda_s^{\alpha_n})$ are completely defined by the sets of numbers (λ, α).

Direct problem

Step 3. The study is performed in two ways depending on whether we proceed from the class of the matrices or from a particular fixed matrix.

a) When the class of the matrices $\{A\}$ is specified, in view of the arbitrariness of the nonsingular matrix V from (1.4.3$'$.I) and by putting the first multiplier of the scalar product to zero, we obtain a SLAE with respect to the vector B_q or B_n. Let there be a vector B_q, then we have

$$C^{(q)}B_q = d, \qquad (1.4.3'.\text{Ia})$$

where $d = \mu C^{(n)}B_n$, $C^{(p)} = (C_s^{(p)})$. Henceforward, the matrices of the $C^{(p)}$ type consist of row vectors. Let us write the condition of the last system's solvability

$$\operatorname{rang} C^{(q)} = \operatorname{rang}(C^{(q)}, d). \qquad (1.5.1)$$

Thus, two rows of the boundary matrix B are defined, i. e. the vector B_q, calculated from (1.4.3′.Ia) and the vector B_n specified in accordance with the condition (1.5.1). The remaining (arbitrary) elements b_{ij} will be selected so as to fulfill the key condition $\det B \neq 0$.

Permissible set of matrices V

As indicated in Lemma 1.4.2, the problem construction depends on the matrices V, so we will describe a set of permissible matrices. Substitution of the sets of eigenvalues (λ) and "exponents" α, of the obtained vectors B_q, B_n as well as of the constant μ, into (1.4.3′.I) yields the linear system $DV^\top = 0$ of N equations with N^2 unknown v_{ij}. This is a system with respect to the vectors V_s, therefore, it is sufficient to solve "eigen" equation of the type (1.2.11) for each fixed index $\bar{s}$. Since each time one equation with N desired values is sought here, obviously there exists a whole family of solutions. We choose any of them provided that there holds the key requirement of the sought matrix nondegeneracy ($\det V^\top = 0$).

b) For the fixed matrix V_φ to the bilinear system (1.4.3′.I) there corresponds the linear system

$$C_V B_q = d_V, \qquad (1.4.3\text{"}'.\text{Ib})$$

completely defined by the above specified sets of numbers (λ, α) and the matrix V_φ. This variant is equivalent to the choice of one matrix A_φ of the class with the given Jordan form J_A. Further the procedure applied to this system is the same as used in (1.4.3′.Ia).

Step 4. Existence of the inverse problem. For the calculation it is sufficient to know the vectors B_q, B_n, whereas from (1.4.3′.I) we arrive at the system

$$D_B V^\top = \mu D_B^1 V^\top, \qquad (1.4.3'.\text{Ic})$$

or, according to the different notation $D_B(\mu)V^\top = 0$

(see item a for information about the matrix D and about the solution of the corresponding systems).

Step 5. Reconstruction of the problem elements. Using the sets of numbers (λ, α) and the determined matrices V and B we find the matrix $A = V^{-1}JV$ of the ODE system and the boundary matrix $\mathcal{F} = B(\partial)\Phi(x)|_{x=0}$.

Step 6. Verification of the condition of ill-posedness $\det \mathcal{F} = 0$.

Proportionality of the columns with the same numbers

Variant C. We solve the nonlinear matrix equation

$$B(\lambda_q)V_q = \mu B(\lambda_n)V_n \tag{1.4.3.II}$$

using the same linearization algorithm.

We will now give only the special distinctions from the above-mentioned.

Step 1′. If we specify eigenvalues λ_q and λ_n only, we obtain

$$B_j(\lambda_q^{\alpha_j} V_q - \mu\lambda_n^{\alpha_j} V_n) = 0, \qquad j = 1, \ldots, N.$$

Step 2′. Choosing the "exponents" α_{ij} we arrive at the homogeneous bilinear system with the constant μ

$$(C_s^{(q)} V_q - \mu C_s^{(n)} V_n, B_s) = 0, \qquad s = 1, \ldots, N \tag{1.4.3′.II}$$

this time with respect to the pair (V_q, V_n) or the set of vectors B_s. Here vectors $C_s^{(q)} = (\lambda_q^{\alpha_s})$ constitute the matrix $C^{(q)} = (\lambda_q^{\alpha_{ij}})$.

Step 3′. Inverse problem. a) For every specified nonsingular B, we solve the system

$$C^{(q)}V_q = d, \tag{1.4.3′.IIa}$$

where $d = \mu C^{(n)}V_n$. As a result, we determine $V = (v_{ij})$, where the column V_q is found from (1.4.3′.IIa), and the vector V_n is given in accordance with the condition of solvability (1.5.1), and the remaining elements are chosen with allowance for its nonsingularity. Since in this case α_s was specified arbitrarily, the permissible set of the matrices B should be defined more precisely. It is done on the basis of (1.4.3′.II) by the procedure of Step 3a).

b) Let us fix the matrix B_φ. The remaining sought vectors in system (1.4.3′.II) are the vectors V_q, V_n and parameter μ, in particular, it may be rewritten in the form

$$C_{B_\varphi} V_q = d_{B_\varphi}(\mu).$$

Step 4′. Direct problem. It is sufficient to give two column vectors V_q, V_n. Then rewriting system (1.4.3′.II) as $D_V(\mu)B = 0$ we find the elements (b_{ij}).

Remark 1.5.1. For the direct problem, it would be better to use the proportionality of the rows (variant d), and for the inverse problem it would be feasible to use the proportionality of the columns (variant c).

Remark 1.5.2. It is easily seen that the presented algorithm by its nature is a multi-layer cyclic process. The cycle can be repeated starting from any step (level), or, to be more precise, when a chosen set of parameters proves unsatisfactory, we can return to the respective step and continue calculation using a new set.

1.5.1. Illustration of the algorithm ($N = 3$)

d) The rows ($q = 1$, $n = 1$) are proportional. We solve the nonlinear vector equation

$$\sum_{j=1}^{3} (b_{1j}\lambda_S^{\alpha_{1j}} - \mu b_{2j}\lambda_S^{\alpha_{2S}})v_{js} = 0, \qquad s = 1,2,3. \qquad (1.2.12'.\mathrm{I})$$

Step 1. Assuming $\lambda_S = -s$, we have

$$\sum_{j=1}^{3} [b_{1j}(-s)^{\alpha_{1j}} - \mu b_{2j}(-s)^{\alpha_{2j}}]v_{js} = 0, \qquad s = 1,2,3. \qquad (1.2.12'.\mathrm{I})$$

Step 2. Let

$$\alpha = \begin{bmatrix} \alpha_1 \\ \alpha_2 \\ \alpha_3 \end{bmatrix} = \begin{pmatrix} 1 & 0 & 0 \\ 0 & 0 & 1 \\ 0 & 1 & 1 \end{pmatrix},$$

then we find

$$C^{(1)} = \begin{bmatrix} \lambda_1^{\alpha_1} \\ \lambda_2^{\alpha_2} \\ \lambda_3^{\alpha_2} \end{bmatrix} = \begin{bmatrix} -1 & 1 & 1 \\ -2 & 1 & 1 \\ -3 & 1 & 1 \end{bmatrix}, \qquad C^{(2)} = \begin{bmatrix} \lambda_1^{\alpha_{12}} \\ \lambda_2^{\alpha_2} \\ \lambda_3^{\alpha_2} \end{bmatrix} = \begin{bmatrix} 1 & 1 & -1 \\ 1 & 1 & -2 \\ 1 & 1 & -3 \end{bmatrix}.$$

Step 3. a) Direct problem.

$$C^{(1)}B_1 = d, \qquad (1.4.3'\mathrm{Ia})$$

where $d = \mu C^{(2)}B_2$. Let $\mu = 1$, $B_{\overline{2}}(-1,0,1)$, then $d = (-2,-3,-4)$ and $\operatorname{rang} C^{(1)} = \operatorname{rang}(C^{(1)}, d) = 2$. After we find $B_1 = (1,2,-3)$ from (1.4.3$'$.Ia) and specify the vector B_3 conforming to the requirement $\det B = 1$ the sought matrix can be written:

$$B(\lambda) = \begin{pmatrix} \lambda & 2 & -3 \\ -1 & 0 & \lambda \\ 2 & -5/2 & \lambda \end{pmatrix}.$$

Permissible set of the matrices V.

In view of the specified set α_{ij} and allowing for the defined vectors B_1, B_2 and constant $\mu = 1$ from (1.4.3′) we arrive at $DV^\top = 0$:

$$\begin{cases} 2v_{21} - 2v_{31} = 0, \\ -v_{12} + 2v_{22} - v_{32} = 0, \\ -2v_{13} + 2v_{23} = 0. \end{cases} \tag{1.4.3′Is}$$

Hence follows the set of solutions:

$$\{v_{11}, v_{21} = v_{31};\ 2v_{22} = v_{12} + v_{32};\ v_{13} = v_{23}, v_{33}\}$$

Choosing v_{ij} such that $\det V = 1$, as a result we find the matrix of the boundary SLAE

$$\mathcal{F}_a = B(\partial)\Phi(x)\big|_{x=0}$$
$$= \begin{bmatrix} \partial & 2 & -3 \\ -1 & 0 & \partial \\ 2 & -(5/2)\partial & \partial \end{bmatrix} \begin{pmatrix} e^{-x} & e^{-2x} & e^{-3x} \\ 2e^{-x} & 0 & e^{-3x} \\ 2e^{-x} & -e^{-2x} & 0 \end{pmatrix}\Bigg|_{x=0}$$
$$= \begin{Vmatrix} -3 & 1 & -1 \\ -3 & 1 & -1 \\ 5 & 4 & 19/2 \end{Vmatrix}.$$

b) If we fix the matrix

$$V = \begin{bmatrix} 1 & 1 & 1 \\ 2 & 1 & 1 \\ 1 & 2 & 1 \end{bmatrix}, \qquad \det V = 1, \qquad V^{-1} = \begin{pmatrix} -1 & 1 & 0 \\ -1 & 0 & 1 \\ 0.3 & -1 & -1 \end{pmatrix}$$

we obtain the system

$$C_V B_1 = d_V. \tag{1.4.3′.Ib}$$

At the same time

$$C_V = \begin{bmatrix} -1 & 2 & 1 \\ -2 & 1 & 2 \\ -3 & 1 & 1 \end{bmatrix}, \qquad \operatorname{rang} C_V = 3, \qquad d_V = \mu \begin{bmatrix} b_{21} + 2b_{22} - b_{23} \\ b_{21} + b_{22} - 4b_{23} \\ b_{21} + b_{22} - 3b_{23} \end{bmatrix}.$$

Whence we calculate the elements b_{1j}

$$b_{11} = -\mu(b_{21} - 4b_{23})/6,$$
$$b_{12} = \mu(b_{21} + 3b_{22} + 2b_{23})/3, \qquad b_{13} = \mu(b_{21} - 10b_{23})/6.$$

Let

$$B(\lambda) = \begin{bmatrix} \lambda & 2 & -3 \\ 1 & 0 & \lambda \\ 0 & -3\lambda/4 & \lambda \end{bmatrix},$$

then

$$A = \begin{pmatrix} -6 & 2 & 1 \\ -5 & 1 & 1 \\ -4 & 2 & 1 \end{pmatrix}, \qquad \mathcal{F}_a = \begin{pmatrix} 0 & -6 & -4 \\ 0 & -3 & -2 \\ 1/2 & -5/2 & -3/4 \end{pmatrix}.$$

Step 4. Inverse problem. Let $B_1 = (1, -1, -1)$ and $B_2 = (2, 2, -1)$, then we find

$$D_B(\mu)V^\top = 0, \qquad D_B(\mu) = \begin{bmatrix} 1 + 2\mu 1 + 2\mu 1 + \mu \\ 2 + 2\mu 1 + 2\mu 1 + 2\mu \\ 3 + 2\mu 1 + 2\mu 1 + 3\mu \end{bmatrix}. \qquad (1.2.12'.\text{Ic})$$

Assuming $\mu = -1/2$, we obtain

$$V = \begin{bmatrix} v_{11} & 0 & v_{13} \\ v_{21} & v_{22} & v_{23} \\ 0 & v_{32} & 4v_{13} \end{bmatrix}.$$

Step 5. Let us reconstruct the elements for the particular case

$$A = V^{-1}JV = \begin{bmatrix} 1 & 0 & -1/4 \\ 0 & 1 & 0 \\ 0 & 0 & 1 \end{bmatrix} \operatorname{diag}(-1,-2,-3) \begin{bmatrix} 1 & 0 & 1/4 \\ 0 & 1 & 0 \\ 0 & 0 & 1 \end{bmatrix}$$

$$= \begin{bmatrix} -1 & 0 & 1/2 \\ 0 & -2 & 0 \\ 0 & 0 & -3 \end{bmatrix},$$

$$\mathcal{F} = \begin{pmatrix} \partial & -1 & -1 \\ 2 & 2 & -\partial \\ 1 & \partial & -(1/4)\partial \end{pmatrix} \begin{pmatrix} e^{-x} & 0 & e^{-3x}/4 \\ 0 & e^{-2x} & 0 \\ 0 & 0 & e^{-3x} \end{pmatrix} \Bigg|_{x=0}$$

$$= \begin{Vmatrix} -1 & -1 & -7/4 \\ 2 & 2 & 7/2 \\ -1 & -2 & 1 \end{Vmatrix}.$$

Variant c. The columns ($q = 1$, $n = 1$) are proportional: $B(\lambda_1)V_1 = \mu B(\lambda_2)V_2$.

Step 1′. Let $\lambda_1 = -1$, $\lambda_2 = -2$, then we obtain

$$b_{i1}[(-1)^{\alpha_{i1}}v_{11} - \mu(-2)^{\alpha_{i1}}v_{12}] + b_{i2}[(-1)^{\alpha_{i2}}v_{21} - \mu(-2)^{\alpha_{i2}}v_{22}] + b_{i3}[(-1)^{\alpha_{i3}}v_{31} - \mu(-2)^{\alpha_{i3}}v_{32}] = 0,$$

where $i = 1, 2, 3$.

Step 2′. Let α_{ij} be fixed, then with respect to the pair of the columns V_1, V_2 and vectors B_s we arrive at the system with the matrices

$$C^{(1)} = (\lambda_1^{\alpha_{ij}}) = \begin{bmatrix} -1 & 1 & 1 \\ 1 & 1 & -1 \\ 1 & -1 & -1 \end{bmatrix}, \qquad C^{(2)} = (\lambda_2^{\alpha_{ij}}) = \begin{bmatrix} -2 & 1 & 1 \\ 1 & 1 & -2 \\ 1 & -2 & -2 \end{bmatrix}. \tag{1.4.3′.II}$$

Step 3′. Inverse problem. a) We solve the system

$$C^{(1)}V_1 = d, \tag{1.4.3′.IIa}$$

where $d = \mu C^{(2)}V_2$. Let $\mu = 1$ and $V_2 = (1, 0, -1)$, then $d = (-3, 3, 3)$, $\operatorname{rang} C^{(1)} = \operatorname{rang}(C^{(1)}, d) = 2$. On finding $V_1 = (2, 0, -1)^\top$, we obtain

$$V = \begin{pmatrix} 2 & 1 & 0 \\ 0 & 0 & 1 \\ -1 & -1 & 0 \end{pmatrix}, \qquad \det V = 1.$$

In order to define the class of permissible B let us write out system (1.4.3$'$.II)

$$\begin{cases} b_{11}\cdot 0 + b_{12}\cdot 0 + b_{13}\cdot 0 = 0, \\ b_{21}\cdot 1 + b_{22}\cdot 0 - b_{23} = 0, \\ b_{31} + b_{32}\cdot 0 - b_{33} = 0. \end{cases}$$

Hence we obtain

$$B = \{(b_{11}, b_{12}, b_{13});\ (b_{21}, b_{22}, b_{21});\ (b_{31}, b_{32}, b_{31})\}.$$

b) By fixing the matrix

$$B = \begin{bmatrix} 1 & 3 & 1 \\ 2 & 0 & 1 \\ 0 & 1 & 0 \end{bmatrix},$$

we arrive at the system

$$C_B V_1 = d_{V_2},$$

where

$$C_B = \begin{bmatrix} -1 & 3 & 1 \\ 2 & 0 & -1 \\ 0 & -1 & 0 \end{bmatrix} = B(\lambda_1),$$

$$d_{V_2} = \mu B(\lambda_2) V_2 = \mu \begin{bmatrix} -2 & 3 & 1 \\ 2 & 0 & -2 \\ 0 & -2 & 0 \end{bmatrix} \begin{bmatrix} v_{12} \\ v_{22} \\ v_{32} \end{bmatrix}.$$

Whence it follows:

$$v_{11} = -\mu(3v_{22} + v_{32}), \quad v_{21} = 2\mu v_{22}, \quad v_{31} = -2\mu(v_{12} + 3v_{22}).$$

Step 4$'$. Direct problem. By specifying the vectors $V_1 = (1, 2, 3)^\top$, $V_2 = (0, 1, 2)^\top$ we arrive at the system $D_V(\mu)B = 0$, with the matrix

$$D_V(\mu) = \begin{pmatrix} -1 & 2-\mu & 3-2\mu \\ 1 & 2-\mu & -3+4\mu \\ -1 & -2+2\mu & -3+4\mu \end{pmatrix}.$$

If $\mu = 1$, we obtain a boundary matrix in the form

$$B = \begin{pmatrix} b_{11} & b_{12} & -b_{11} \\ -5b_{23} & b_{22} & b_{23} \\ b_{31} & b_{32} & -(b_{31} + 2b_{32})/5 \end{pmatrix}.$$

Chapter 2.

Parabolic problems

2.1. PARABOLIC SYSTEMS

In mathematical publications different kinds of parabolic systems are considered. Let us name the most important ones: those are parabolic systems in Petrovskii's sense, parabolic systems in Shirota's sense, and parabolic systems in Solonnikov's sense.

Petrovskii (1938) introduced a wide class of parabolic systems, linear partial differential equations, generalizing the equation of thermal conductivity. Its definition proved to be very successful because many inherent properties of the equation solution were retained for such systems, both linear and nonlinear. Then parabolic systems in Shirota's sense were studied, which were first-order systems (Shirota, 1998). Systems of differential equations studied by Solonnikov (1965) proved to be more general than the previous two kinds of systems.

2.1.1. Types of parabolic systems

2.1.1.1. Parabolic system in Petrovskii's sense

Equation. Let us take a cylinder $D = \Omega \times [0,T]$ in $\mathbb{R}^{n+1}$ with the base $\Omega \in \mathbb{R}^n$ and the smooth boundary $S = \partial\Omega$. Let us denote the lateral surface by $\Gamma \equiv \partial D = S \times [0,T]$, points of the domain Ω by $x = (x_1, \dots, x_n)$, points of the segment $[0,T]$ by t, the points of the cylinder D by (t, x).

Let there be specified a linear differential operator $\mathcal{L}(t, x; \partial_t, \partial_x)$ of an arbitrary order with complex coefficients, such that at every point $(t, x) \in D$ the function $L(t, x; p, i\xi)$ will be a polynomial in p and ξ when $\xi \in \mathbb{R}^n$ and p is a scalar complex parameter. The degree of the polynomial $L(t, x; p\lambda^{2b}, i\xi\lambda)$ in λ is equal to $2br$ where b and r are integers. The sum of all the terms, for which the homogeneity degree is $2br$

$$L_0(t, x; p\lambda^{2b}, i\xi\lambda) = \lambda^{2br} L_0(t, x; p, i\xi) \tag{2.1.1}$$

is called the principal part of the polynomial L.

Definition 2.1.1. The operator $\mathcal{L}(t, x; \partial_t, \partial_x)$ is called $2b$-parabolic at the point (t, x), if the p_s-roots of the polynomial $L_0(t, x; p, i\xi)$ for every real $\xi \in \mathbb{R}^n$ satisfies the inequality

$$\operatorname{Re} p_s \leq -\delta|\xi|^{2b}, \qquad \delta > 0. \tag{2.1.2}$$

Hence it follows that for parabolic operators a single differentiation with respect to t is equivalent to $2b$-fold differentiation with respect to x (the derivative ∂_t is said to have the weight $2b$). This weight must necessarily be an even number, otherwise the condition of parabolicity (2.1.2) will no longer hold for every real ξ.

For the parabolic operator $\mathcal{L}$, the numbers b and r are uniquely defined, and r here is the degree of the polynomial $L(t, x; p, i\xi)$ with respect to p, and $2br$ with respect to ξ_j.

System. In the domain D we consider linear systems of N equations containing N unknown functions $(u_1, \ldots, u_N)$ of the form

$$\sum_{j=1}^{N} \ell_{kj}(t, x, \partial_t, \partial_x) u_j(t, x) = f_k(t, x), \qquad k = 1, \ldots, N. \tag{2.1.3}$$

Here ℓ_{kj} are polynomials over the last two arguments; $f_k(t, x)$ are given functions.

System (2.1.3) can conveniently be rewritten in the matrix-vector form

$$\bar{\mathcal{L}} u = f. \tag{2.1.3$'$}$$

Definition 2.1.2. The matrix operator $\bar{\mathcal{L}}(t, x; \partial_t, \partial_x)$ is called parabolic in Petrovskii's sense if:

1) the operator $\mathcal{L}(t, x; \partial_t, \partial_x) = \det \bar{\mathcal{L}}(t, x; \partial_t, \partial_x)$ is $2b$ parabolic;

2) the degree of the polynomial $\ell_{kj}(t,x,p\lambda^{2b},i\xi\lambda)$ by λ does not exceed $2bn_j$ and there is a representation

$$\ell_{kj}(t,x,p,i\xi) = \delta_{kj}p^{n_j} + \ell'_{kj}(t,x;p,i\xi),$$

where δ_{kj} is the Kronecker symbol; the degree of the polynomial ℓ'_{kj} by p is strictly lower than n_j.

The latter implies that the system is resolved with respect to higher derivatives $\partial_t^{n_j}u_j$ and has the following form

$$\partial_t^{n_j}u_j(t,x) = \sum_{s=1}^{N}\sum_{2bk_0+|k|\le 2bn_s} A_{is}^{(k_0,k)}(t,x)\partial_t^{k_0}\partial_x^k u_s(t,x), \qquad j=1,\dots,N. \tag{2.1.4}$$

The quantity $2b\max_j n_j$ determines the order of the system. The principal parts $\bar{L}_0$ and $\bar{L}$ of the matrix $\bar{L}$ and polynomial $L(t,x_j,p,i\xi)$ are respectively composed of the principal parts ℓ_{kj}^0 of the polynomial ℓ_{kj}, found by the principle (2.1.1):

$$L_0(t,x,p,i\xi) = \det \bar{L}_0(t,x,p,i\xi).$$

An important and much-studied particular variant of parabolic systems in Petrovskii's sense are systems, for which every $n_j = 1$, which therefore assumes the form

$$\mathcal{L}u \equiv u_t + A(t,x,\partial_x)u = f(t,x), \tag{2.1.4$'$}$$

where the operator A is of the order $2b$ with respect to the derivatives ∂x_j.

Eidelmann (1964) proved that if parabolicity in Petrovskii's sense is retained, every higher-order system (2.1.4) can be reduced to the first-order system (2.1.4′). Therefore, it is sufficient further to study parabolic systems in Petrovskii's sense of the type (2.1.4′).

2.1.1.2. Parabolic system in Shirota's sense

As it is known, the homogeneity degree of polynomials ℓ_{kj}^0 equal to $2br_j$ does not depend on the number k of the equation for parabolic systems in Petrovskii's systems. Therefore, the definition of parabolicity was suggested taking into account also the number k. For hyperbolic systems, such generalization had been performed even earlier by J. Leray, and for elliptic systems by A. Douglas and L. Nirenberg. In their works the order of the operators ℓ_{kj} is determined by the sum $s_k + t_j$, where s_k and t_j are sets of integers. For parabolic systems such a generalization was introduced by Japanese mathematician Shirota (1998).

Definition 2.1.3. A system of the type (1.1.4′) is called parabolic in Shirota's sense if:

1) the operator $\mathcal{L} = \det \bar{\mathcal{L}}$ is $2b$-parabolic;

2) there exist numbers s_k, $k = 1, \ldots, N$, such that the order of the operator A_{kj} does not exceed the value $s_k - s_j + 2b$.

Even though these systems contain only first- and second-order derivatives with respect to t, and Petrovskii's systems (2.1.4) contain the derivatives of an arbitrary order, the latter ones are a subclass of the former ones. As has already been noted, every parabolic system in Petrovskii's sense (2.1.4) can be reduced to system (2.1.4′), then it will necessarily be a parabolic system in Shirota's sense, and the opposite is not always true. For example, the system

$$u_t = a_1 u_{xx} - i a_2 v_{xxx},$$
$$v_t = i a_3 u_x + a_4 v_{xx}$$

for which $b = 1$, $s_1 = 0$, $s_2 = -1$, $t_j = 2 - s_j$ is parabolic in Shorota's sense if for the real a_j the following inequalities hold true

$$a_2 + a_4 > 0, \qquad a_2 a_4 - a_2 a_3 > 0.$$

It will not, however, be parabolic in Petrovskii's sense.

2.1.1.3. Parabolic system in Solonnikov's sense

Still wider class of parabolic systems is defined in the paper by Solonnikov (1965).

Definition 2.1.4. An operator $\bar{L}$ (1.1.3′) is called parabolic in Solonnikov's sense, if:

1) the operator $L = \det \bar{L}$ is $2b$-parabolic in the sense of Definition 1;

2) there exist integers s_k and t_k, $k = 1, \ldots, N$, such that the degree of the polynomial $\ell_{kj}(t, x, p\lambda^{2b}, i\xi\lambda)$ with respect to the parameter λ does not exceed $s_k + t_j$ and in addition,

$$\sum_{k=1}^{N} (s_k + t_k) = 2br,$$

in the case $s_k + t_j < 0$ it is assumed that $\ell_{kj} \equiv 0$.

The principal part $\bar{L}_0$ of the matrix $\bar{L}$ is composed of polynomials being a sum of all the terms ℓ_{kj}, for which the homogeneity condition is satisfied:

$$\ell^0_{kj}(t, x; p\lambda^{2b}, i\xi\lambda) = \lambda^{s_k+t_j}\ell^0_{kj}(t, x; p, i\xi).$$

If the degree of the polynomial $\ell_{kj}(t, x; p\lambda^{2b}, i\xi\lambda)$ is less than $s_k + t_j$, then $\ell_{kj} \equiv 0$. Evidently, the polynomial $L_0 = \det \bar{L}_0$ is the principal part of the polynomial $L(t, x, p, i\xi)$.

2.1.2. Power of parabolic systems

Let us show that parabolic systems in Petrovskii's (2.1.1) and Shirota's (2.1.2) senses are the subclasses of parabolic systems in Solonnikov's sense. According to the definitions, we have

a) Parabolic systems in Solonnikov's sense are such that

   α) $s_j = 0$ for every $j = 1, \ldots, N$ and $t_j = 2bn_j$, $n_j > 0$;

   β) $\ell^0_{ij}(t, x, p, 0) = \delta_{ij}p^{n_j}$.

b) if a parabolic system in Solonnikov's sense is given in the form (1.1.4′), then for diagonal operators we have

$$\ell^0_{jj}(t, x, p\lambda^{2b}, i\xi\lambda) = p\lambda^{2b} + A^0_{jj}(t, x; i\xi\lambda) = \lambda^{s_j+t_j}\ell^0_{jj}(t, x; p, i\xi).$$

   Therefore the last one is a parabolic system in Shirota's sense if

   α') $s_j + t_j = 2b$ ($t_j = 2b - s_j$) for all $j = 1, \ldots, N$;

   β') $\ell_{kj} = \delta_{kj}p + a_{kj}(t, x; i\xi)$,

   where a_{kj} is a polynomials with respect to ξ with a degree not exceeding $S_k + 2b - S_j$.

Remark 2.1.1. In order to restrict freedom in choosing the pair s_k, t_j (since $(s_k + a) + (t_j - a) = s_k + t_j$, a is any constant), we impose condition $\max_k s_k = 0$ fixing the constant a.

System (Solonnikov, 1965)

$$a_1 u_t + a_2 v_t - a_3 \Delta v = 0,$$
$$a_3 v_t - a_2 \Delta u + a_4(v_{tt} + \Delta v) = 0,$$

will be neither parabolic system in Solonnikov's sense, nor in Shirota's sense, because for it

$$b = 1, \quad s_1 = -2, \quad t_1 = 2, \quad s_2 = 0, \quad t_2 = 4, \quad r = 2,$$

but it is will be parabolic system in Solonnikov's sense if

$$a_2^2 + a_3^2 > 0, \qquad a_1 a_4 - a_2 a_3 < 0.$$

In conclusion, let us point out the following important features:

1) the parabolicity definition contains two important criteria: i) Property of the roots of (algebraic) characteristic equation (see (2.1.2) from Definition 2.1.1) (root criterion); ii) certain assumption about the structure of the equations (a certain set of numbers $t_j + s_k$) (structural criterion).

2) The root criterion choose a $2b$-parabolic differential equation from the whole of them, while the distribution over types proceeds on the basis of the second criterion.

3) The root criterion was also generalized and as a result are parabolic systems in Shilov's and Eidelmann's senses.

4) A question whether it is always possible to select numbers s_j and t_j in such a way that the operator $\mathcal{L}$ would be a parabolic system in Solonnikov's sense and whether such a set would be unique was answered by Volevitch (1960): such numbers always exist but cannot always be uniquely defined despite the restrictions imposed by Remark 2.1.1.

2.1.3. Description of the parabolic systems in Petrovskii's sense ($N = 2$, $b = 1$). Examples

Consider the system of the general form

$$\begin{aligned} u_t &= a_1 u_{xx} - i a_2 u_{xy} + a_3 u_{yy} - a_4 v_{xx} - i a_5 v_{xy} - a_6 v_{yy}, \\ v_t &= -b_1 u_{xx} - i b_2 u_{xy} + b_3 u_{yy} - b_4 v_{xx} - i a_2 v_{xy} - b_6 v_{yy}, \end{aligned} \tag{2.1.5}$$

where a_j, b_j are real numbers.

As can easily be seen, the characteristic equation of this system can be written in the form:

$$p^2 + \alpha p(\tau^2 + \xi^2) + \beta(\tau^2 + \xi^2)^2 = 0. \tag{2.1.6}$$

The inequalities

$$\alpha > 0, \qquad \beta > 0 \tag{2.1.7}$$

also ensure the parabolicity of the system ($\operatorname{Re} p_{1,2} < -\delta^2\xi^2$, $\delta > 0$) and the existence of τ^+ roots such that $\operatorname{Im} \tau^+ > 0$:

$$\tau^+_{1,2} = i\sqrt{\xi^2 + p(\alpha/(2\beta) \pm \sigma)}, \qquad \sigma = \sqrt{\alpha^2/(4\beta^2) - 1/\beta}. \tag{2.1.8}$$

Thus, τ-roots are simple if $\sigma > 0$ and they are multiple when $\sigma = 0$ ($\alpha^2 = 4\beta$). Thereby, we have subdivided parabolic system in Petrovskii's sense into two classes. This will be called subdivision of the systems on the sense basis and denoted by $S1$ ($\sigma > 0$) and $S2$ ($\sigma = 0$).

Let us justify inequalities (2.1.7), namely, if $\alpha \leq 0$ the parabolicity condition will be violated, and if $\beta \leq 0$, then τ-roots will be lost.

Rewriting the left-hand side of the characteristic equation as a multinomial over τ we obtain ODE by x with respect to $u(p, x, \xi)$:

$$u^{(\mathrm{IV})} - (\alpha p/\beta + 2\xi^2)u'' + (p^2/\beta + \alpha p\xi^2/\beta + \xi^4)u(p, x, \xi) = 0. \tag{2.1.9}$$

The function $v(p, x, \xi)$ is obtained from the parameterized system of ODE after its canonization, which is always possible, because $\beta = -(a_1b_4 + a_4b_1) > 0$:

$$\begin{aligned} \beta u'' &= (a_2b_4 - a_4b_2)\xi u' + (a_2b_4 + a_5b_4)\xi v' \\ &\quad + [b_4p + (a_3b_4 - a_4b_3)\xi^2]u + [a_4p + (a_6b_4 - a_4b_6)\xi^2]v, \\ \beta v'' &= -(a_2b_1 + a_1b_2)\xi u' + (a_1b_2 + a_5b_1)\xi v' \\ &\quad + [b_1p + (a_1b_3 + a_3b_1)\xi^2]u + [a_1p - (a_1b_6 + a_6b_1)\xi^2]v. \end{aligned}$$

Further, let the following relations between the coefficients be true:

$$a_2b_4 + a_5b_4 = 0, \qquad a_4p - (a_4b_6 - a_4b_6)\xi^2 \neq 0, \tag{2.1.10}$$

then the first equality assumes the form

$$\begin{aligned} &[a_4p + (a_6b_4 - a_4b_6)\xi^2]v \\ &\qquad = \beta u'' + (a_4b_2 - a_2b_4)\xi u' + [b_4p + (a_3b_4 - a_4b_3)\xi^2]u, \end{aligned} \tag{2.1.11}$$

from which the function v can be readily found.

If in addition to (2.1.10) there holds the equation

$$a_4b_2 = a_2b_4, \tag{2.1.12}$$

the form is simplified

$$[a_4p + (a_6b_4 - a_4b_6)\xi^2]v = \beta u'' + [b_4p + (a_3b_4 - a_4b_3)\xi^2]u. \qquad (2.1.13)$$

Now let us find out whether there are real numbers a_j, b_j satisfying all those conditions, i. e. whether the set of PS is non-empty.

Let us expand the determinant of the parameterized ($\partial_t \sim p$, $\partial_x \sim i\tau$, $\partial_y \sim i\xi$) algebraic system (Solonnikov, 1965):

$$\begin{vmatrix} p + a_1\tau^2 - ia_2\tau\xi + a_3\xi^2 & -a_4\tau^2 - ia_5\tau\xi - a_6\xi^2 \\ -b_1\tau^2 - ib_2\tau\xi + b_3\xi^2 & p - b_4\tau^2 + ia_2\tau\xi - b_6\xi^2 \end{vmatrix}$$
$$= p^2 + [(a_1 - b_4)\tau^2 + (a_3 - b_6)\xi^2]p - (a_1b_4 + a_4b_1)\tau^4$$
$$+ i\tau\xi[(a_2b_4 - b_2a_4 + a_1a_2 - a_5b_1)\tau^2 + (a_2b_6 - b_2a_6 + a_2a_3 + a_5b_3)\xi^2]$$
$$+ (a_2^2 - a_3b_4 - a_1b_6 + b_3a_4 + a_5b_2 - a_6b_1)\tau^2\xi^2 - (a_3b_6 - a_6b_3)\xi^4.$$

It must coincide with the characteristic equation (2.1.6). Their comparison yields five equalities:

$$a_1 - a_3 + b_6 - b_4 = 0, \qquad (2.1.14.1)$$
$$a_1b_4 + a_4b_1 = a_3b_6 - a_6b_3, \qquad (2.1.14.2)$$
$$a_4b_3 - a_3b_4 + a_2^2 + a_5b_2 - a_6b_1 - a_1b_6 + 2(a_1b_4 + a_4b_1) = 0, \qquad (2.1.14.3)$$
$$a_2(a_1 + b_4) = a_5b_1 + a_4b_2, \qquad (2.1.14.4)$$
$$a_2(a_3 + b_6) = a_6b_2 - a_5b_3. \qquad (2.1.14.5)$$

Besides

$$\alpha = a_1 - b_4 > 0, \qquad \beta = -(a_1b_4 + a_4b_1) > 0. \qquad (2.1.7')$$

In order to motivate the validity of the above-mentioned requirements we present the values of the coefficients from Solonnikov (1965) counter-example

$$a_1 = b_1 = -b_4 = 1, \quad a_2 = -b_2 = 2a, \quad a_3 = -(a^2 + \mu),$$
$$b_3 = a^2, \quad a_4 = a_5 = 0, \quad a_6 = 4a^2, \quad b_6 = -(1 + 2a^2),$$

where $a = \sqrt{1 + \mu}$, $\mu > 0$.

Remark 2.1.2. The system of equalities (2.1.14) identify parabolic systems in Petrovskii's sense and relations (2.1.10), (2.1.12) subdivide them into groups on the form basis, namely, if there holds only (2.1.10), then there are mixed derivatives in system (2.1.5); and if in addition (2.1.12) holds, then there are no mixed derivatives in system (2.1.5). These groups are denoted respectively by letters Φ1 and Φ2.

Remark 2.1.3. In order to pass to the vector $y = (y_1, \dots, y_m)$ it is sufficient to substitute the quantity ξ^2 by $|\xi|^2 = \xi_1^2 + \cdots + \xi_m^2$. All the considerations and results are still valid.

Examples

Let us analyze relations (2.1.10), (2.1.12), and (2.1.14) in order to separate out of (2.1.5) systems of the classes of interest to us, i. e. $S1$ and $S2$ and of the groups Φ1 and Φ2.

The equality in (2.1.10) holds, in particular, if at least one of the following conditions holds true:

$$a_4 = a_5 = 0, \quad a_2 = a_4 = 0, \quad a_2 = a_5 = 0, \quad a_4 = b_4 = 0. \tag{2.1.15}$$

If, in addition, we assume that holds at least one of the conditions

$$a_4 = a_2 = 0, \quad \cup a_2 = b_2 = 0, \quad \cup b_2 = b_4 = 0, \tag{2.1.16}$$

then equality (2.1.16) also holds. The last of (2.1.15) cannot be used, because it yields $\beta = 0$. Let us consider the remaining variants.

1. Let $a_4 = a_5 = 0$, then it follows from (2.1.14.4) that $b_4 = -a_1$ or $a_2 = 0$. Let us study them separately:

a) Substitute $b_4 = -a_1$ into (2.1.14.1), then $b_6 = a_3 - 2a_1$, now let us substitute this value into (2.1.14.5) and (2.1.14.4), which gives us $a_6 b_2 = 2a_2(a_3 - a_1)$ and $a_6 b_3 = (a_3 - a_1)^2$ respectively. As a result, equality (2.1.14.3) assumes the form $a_2^2 = a_6 b_1$. Thus, all b_j are determined by a_k, $k = 1, \dots, 6$.

b) Let $a_2 = 0$, then (2.1.14.5) implies $b_2 = 0$ (thereby condition (2.1.16) is satisfied). All the rest we rewrite as a SLAE

$$\begin{aligned} b_4 - b_6 &= a_1 - a_3, \\ a_6 b_3 + a_1 b_4 - a_3 b_6 &= 0, \\ (2a_1 - a_3) b_4 - a_1 b_6 + &= a_6 b_1. \end{aligned}$$

Hence we obtain:

$$\begin{aligned} b_3 &= (a_1 - a_3)^2 / a_6 - b_1, \\ b_4 &= a_6 b_1 / (a_1 - a_3) - a_1, \\ b_6 &= a_6 b_1 / (a_1 - a_3) - 2a_1 + a_3. \end{aligned}$$

Obviously, $a_1 \neq a_3$, $a_6 \neq 0$. Thus we have

a) $\alpha = 2a_1$ and $\beta = a_1^2$ $(\sigma = 0)$;

b) $\alpha = a_6 b_1/(a_3 - a_1) + 2a_1$, $\beta = a_1 a_6 b_1/(a_3 - a_1) + a_1^2$ (if $b_1 = 0$, then $\sigma = 0$).

Let us show one system from each group $\Phi 1$ and $\Phi 2$ belonging to the class $S2$ $(\sigma = 0)$. Assuming $b_1 = 1$ in a), we arrive at

$$\begin{aligned} u_t &= a_1 u_{xx} - i a_2 u_{xy} + a_3 u_{yy} - a_2^2 v_{yy}, \\ v_t &= -u_{xx} - 2i\delta_1 u_{xy} + \delta_1^2 u_{yy} + a_1 \Delta v + i a_2 v_{xy} - a_2 \delta_1 v_{yy}, \end{aligned} \tag{2.1.5.1a}$$

where $\delta_1 = (a_3 - a_1)/a_2$.

From b) we have

$$\begin{aligned} u_t &= a_1 u_{xx} + a_3 u_{yy} - a_6 v_{yy}, \\ v_t &= b_3 u_{yy} + a_1 v_{xx} + (2a_1 - a_3) v_{yy}, \end{aligned} \tag{2.1.5.1b}$$

where $b_3 = (a_1 - a_3)^2/a_6$.

The coefficients a_j can be assigned values from Solonnikov's example.

2. Let $a_2 = 0$, $b_4 = 0$, then equality (2.1.14.1) assumes the form $b_6 = a_3 - a_1$. We consider two independent cases

a) First from the equalities

$$\begin{aligned} a_4 b_1 + a_6 b_3 &= a_3(a_3 - a_1), \\ a_5 b_1 + a_4 b_2 &= 0, \\ a_6 b_2 - a_5 b_3 &= 0 \end{aligned}$$

we determine the values b_j:

$$b_1 = -a_4 \delta_2, \quad b_2 = a_5 \delta_2, \quad b_3 = a_6 \delta_2, \tag{2.1.17}$$

where $\delta_2 = a_3(a_3 - a_1)/(a_6^2 - a_4^2)$.

Then we substitute them into (2.1.14.3)

$$a_1(a_6^2 - a_4^2) = a_3(a_5^2 - 2a_4^2 + 2a_4 a_6). \tag{2.1.18}$$

b) The last two equalities of system (2.1.14) yield $a_5 = b_2 = 0$ that is equivalent to condition (2.1.16). We find from the remaining equalities (2.1.14) that

$$b_1 = \frac{a_3 - a_1}{a_4 - a_6}\left(a_3 + a_6 \frac{a_3 - a_1}{a_4 - a_6}\right), \quad b_3 = \frac{(a_3 - a_1)^2}{a_4 - a_6} - b_1. \tag{2.1.19}$$

If we require fulfillment of inequalities

$$\alpha = a_1 > 0, \qquad \sigma > 0, \qquad \delta_2 > 0$$

then $\beta = a_4^2\delta_2 > 0$. Thereby, we can write systems of interest for us:

$$\begin{aligned} u_t &= a_1 u_{xx} + a_3 u_{yy} - a_4 v_{xx} - i a_5 v_{xy} - a_6 v_{yy}, \\ v_t &= a_4\delta_2 u_{xx} - i a_5 \delta_2 u_{xy} + a_6 \delta_2 u_{yy} + (a_1 - a_3) v_{yy}, \end{aligned} \tag{2.1.5.2a}$$

where a_j are connected by relation (2.1.18),

$$\begin{aligned} u_t &= a_1 u_{xx} + a_3 u_{yy} - a_4 v_{xx} - a_6 v_{yy}, \\ v_t &= -b_1 u_{xx} + b_3 u_{yy} + (a_1 - a_3) v_{yy}, \end{aligned} \tag{2.1.5.2b}$$

where b_1, b_3 are calculated by (2.1.19).

It is easily seen that these systems belong to the class $S1$ ($\sigma > 0$) and belong to the groups $\Phi 1$ and $\Phi 2$ respectively.

2.2. BOUNDARY-VALUE PROBLEMS

2.2.1. General aspects. Mixed problem

Assume that in the cylinder $D = \Omega \otimes [0, T]$, $x \in \Omega \in \mathbb{R}^n$, $0 \le t \le T \le +\infty$, the domain Ω can be bounded or unbounded, as well as its boundary $S = \partial\Omega$.

In initial boundary-value (mixed) problems, both initial and boundary conditions can be given by arbitrary linear differential operators, but they must satisfy two types of requirements:

— connected with the property of the system itself, i. e., quantitative factor, it specifies the number of specifying equalities (number of lines in the data-carrying operator);

— related to the problem well-posedness; it is the qualitative factor of an algebraic character, which is called complementing condition.

For boundary operators this algebraic condition is an analogue of the well-known Lopatinskii condition. The operators responsible for initial data also satisfy the condition whose form is similar to complementing condition, although stronger. For instance, initial data cannot always be reduced to specifying separate functions u_k or their time derivatives, i. e., to the Cauchy form we are accustomed to.

Note the following important facts

1) the quantitative factor must necessarily be satisfied for every problem, in other words, it is the necessary condition for correct (reasonable) problem statement;

2) the validity of a complementing condition implies a unique solvability of the problem; and its (quality factor) violation leads to ill-posedness, i. e., either the problem has a family of solutions, or the problem is solvable not for every right-hand side of the initial data;

3) three types of problems can be formulated:

   I. Cauchy problem (only initial condition is given). It is always well-posed.

   II. Boundary-value problem (without initial condition). Complementing condition is verified.

   III. Mixed problem (initial and boundary conditions are specified). Complementing condition are specified both for initial and boundary operators.

Consequently, the last two types may contain ill-posed problems arising when complementing condition is violated.

Problem. Find the vector $u = (u_1, u_2, \dots, u_N)$ in the domain D, for which the following equalities hold true

$$\mathcal{L}u = f \sim \sum_{j=1}^{N} \ell_{kj}(t, x; \partial_t, \partial_x) u_j(t, x) = f_k(t, x),$$
$$k = 1, \dots, N, \qquad (t, x) \in D, \tag{2.2.1}$$

$$Cu\big|_{t=0} = \varphi \sim \sum_{j=1}^{N} C_{\alpha j}(x; \partial_t, \partial_x) u_j(0, x) = \varphi_\alpha(x),$$
$$\alpha = 1, \dots, r, \qquad x \in \Omega, \tag{2.2.2}$$

$$Bu\big|_{\Gamma} = \Phi \sim \sum_{j=1}^{N} B_{qj}(t, x; \partial_t, \partial_x) u_j(t, x)|_\Gamma = \Phi_q(t, x),$$
$$q = 1, \dots, br, \qquad (t, x) \in \Gamma. \tag{2.2.3}$$

At the same time for system (2.2.6) the parabolicity condition is retained. We will use the following properties of such systems.

Lemma 2.2.1. *The polynomial $L(t,x;p,i(\xi+\nu\tau))$ as the function τ has br roots τ_s^+, $\operatorname{Im}\tau_s^+>0$ and br roots τ_s^-, $\operatorname{Im}\tau_s^-<0$, if only the following inequalities hold*

$$\operatorname{Re}p\geq-\delta_1\xi^{2b},\qquad |p|^2+|\xi|^{4b}>0,\qquad \delta>\delta_1>0,\tag{2.2.4}$$

where $\nu(x)=(\nu_1,\dots,\nu_n)$ is a unit vector of the inner normal to S at the point x, and $\xi(x)$ is the vector lying in the tangential plane to S at the same point x.

For the matrices B and C, the respective complementing conditions are verified.

The principal part of the q-th line of the boundary matrix B is obtained as a sum of the terms of the polynomial B_{qj} satisfying the homogeneity condition

$$B_{qj}^0(t,x_jp\tau^{2b},i\xi x)=\tau^{\sigma_q+t_j}B_{qj}^0(t,x_jp,i\xi),$$

then $B_0(t,x;\rho,i\xi)=(B_{qj}^0)$ is the principal part of the matrix B.

The sum $C_{\alpha j}^0$ of all the terms of the polynomial $C_{\alpha j}$ satisfying the homogeneity condition

$$C_{\alpha j}^0(x;p\lambda^{2b},i\xi\lambda)=\lambda^{\rho_\alpha+t_j}C_{\alpha j}^0(x;\rho,i\xi)$$

is called the principal part of the matrix C. Clearly, the parameters t_j, σ_q, and ρ_α are integers and all $\rho_\alpha\leq-2b$ (are negative).

2.2.2. Unique solvability of the Cauchy problem

Let us consider in the halfspace $D_{n+1}=\{t>0,\ x\in\mathbb{R}^n\}$ the problem constructed from the principal parts of the operators $\mathcal{L}$ and C with constant coefficients

$$\bar{\mathcal{L}}_0(\partial_t,\partial_x)U(t,x)=0,\qquad C_0(\partial_t,\partial_x)U(0,x)=\varphi(x).\tag{2.2.5}$$

In (2.2.5) the Fourier transform by x yields an analogous problem for a system of ODE of higher order

$$\bar{\mathcal{L}}_0(d_t,i\xi)\tilde{u}(t,\xi)=0,\qquad C_0(d_t,i\xi)\tilde{u}(0,\xi)=\tilde{\varphi}(\xi).$$

Since by virtue of the initial system parabolicity the characteristic polynomial $L_0(p,i\xi)=0$ has $r^-=r$ roots p_s^- with the negative real part and not a single root with a positive term, the requirement $|\tilde{u}|\to0$ as

$t \to +\infty$ is automatically satisfied for every solution of the homogenous system $\bar{L}(d_t, i\xi)\tilde{u} = 0$. In addition, it is proved by Solonnikov (1965) that for the matrix $C_0(p, i\xi)$ from (2.2.5) complementing condition is satisfied for any real ξ. This implies that the Cauchy problem is always uniquely solvable.

2.2.3. Problem without initial data

In the half spaces $D_{n+2} = (x > 0,\ t, y \in \mathbb{R}^{n+1},\ n \geq 0)$ the solution of a parabolic system in Petrovskii's sense of the type (2.1.4′) is sought with constant coefficients

$$u_t = A(\partial_x, \partial_y)u, \qquad Au = \sum_{j=1}^{N} \sum_{k_0+k=2b} a_{k,k_0} \partial_x^{k_0} \partial_y^{k} u(t, x, y), \tag{2.2.6}$$

satisfying the general boundary condition

$$B(\partial)u(t,0,y) = \sum_{j=1}^{N} \sum_{2bs+\ell_0+\ell \leq r_i} b_{ij}^{(s,\ell_0,\ell)} \partial_t^s \partial_x^{\ell_0} \partial_y^l u_j(t,0,y) = f_i(t,y),$$
$$i = 1, \dots, r, \quad r = bN. \tag{2.2.7}$$

Using integral transforms we obtain the boundary-value problem for ODE of the order $2b$

$$[A(\partial_x, i\xi) - pI]v(x, p, \xi) = 0, \quad x \geq 0,$$
$$B(\partial_x, p, i\xi)v(0, p, \xi) = \tilde{f}(p, \xi). \tag{2.2.8}$$

The implication of this transform is that this problem should be assumed to be solvable

$$\text{for all real } \xi \in \mathbb{R}^n \text{ and complex } p = a_0 + ip_1,\ a_0 > 0,\ p_1 \in \mathbb{R}. \tag{2.2.9}$$

Consequently, first of all the system solution should be used decreasing at $x \to \infty$. According to Lemma 2.2.1, there exist $r = bN$ linearly independent bounded solutions

$$v_m(x; p, \xi) = P_m(x; p, \xi)\ell^{\lambda_m x},$$
$$P_m(x_j p, \xi) = \sum_{S=0}^{k_m} q_S^m(p, \xi)x^S, \qquad \sum_m k_m = bN, \tag{2.2.10}$$

where $\lambda_m = i\tau_m^+$, $\operatorname{Re}\lambda_m < 0$, k_m is the multiplicity of the root τ_m^+.

We reduce system (2.2.8) to canonical form, which is always possible due to the parabolicity (2.2.6) in Petrovskii's sense, and then using a standard procedure (denoting all the lower-order derivatives by new unknown functions y_k) we obtain the problem for a normal first-order system:

$$\dot{y} = Ay(x), \quad x > 0; \qquad My(0) = \varphi. \tag{2.2.11}$$

Here $A = A(p, \xi)$ is a quadratic matrix of the order $2r$, M is a rectangular matrix of size $r \times 2r$.

Recall that

1) the matrix A has eigenvalues of two types

$$\begin{aligned} &\operatorname{Re}\lambda_j(p,\xi) < 0, \quad j = 1, \dots, r \\ \text{and} \quad &\operatorname{Re}\lambda_s(p,\xi) > 0, \quad s = 1, \dots, r; \end{aligned} \tag{2.2.12}$$

2) if the order of the derivative d_x in the operator B exceeds or is equal to the number $2b$, then $M \equiv M(d_x)$ will be differential, and if the same order is lower than $2b$, then M is a scalar. In the both cases its elements depend on the parameters p and ξ.

Taking into account the relation between the ODE systems of the higher and first order we write the matrix of interest for us:

$$\tilde{\Phi}(x) = [Z(x), Z'(x), \dots, Z^{(b-1)}(x)]^\top. \tag{2.2.13}$$

Here, the columns of the matrix $Z(x)$ of size $N \times r$ are formed from linearly independent bounded solutions (2.2.10). According to property (2.2.12) we zero exactly half of the arbitrary quantities $C(p,\xi)$ from the representation $y = \Phi(x)C$ of the system general solution, and the remaining half is found from boundary data (2.2.11)

$$\mathcal{F}C = \varphi, \quad C(p,\xi) = (C_1, \dots, C_r), \quad \mathcal{F} = M(d_x)\tilde{\Phi}(x)\big|_{x=0}. \tag{2.2.14}$$

Therefore, verification of complementing condition is reduced to the investigation of the matrix $\mathcal{F}(p,\xi)$, namely, if it is nondegenerate for $\operatorname{Re} p > 0$, the complementing condition is satisfied, and otherwise it is violated.

2.3. THE FIRST BOUNDARY-VALUE PROBLEM

2.3.1. Parabolic system in Petrovskii's sense and Eidelmann's hypothesis

1. Eidelmann (1964) used the following assumption: for every parabolic system in Petrovskii's sense the boundary matrix B, which represents the first boundary-value problem

$$\partial_x^s u_j(t,0,y) = f_{sj}(t,y), \qquad s = 0,1,\dots,b-1, \quad j = 1,\dots,N \qquad (2.2.7')$$

satisfies the complementing condition.

No direct proof of this statement (Eidelmann's hypothesis) has been given either by Solonnikov (1965) or Eidelmann (1964). There is only a reference to the results obtained by Mikhailov (1959) concerning the solvability of such problems based on the theory about the generalized heat potentials. Later by Solonnikov (1965) a contrary instance is given in the form of a system with a totality of solutions; thereby Solonnikov showed that for the first boundary-value problem (2.2.7′) completing condition does not always hold true. In this connection, it would be interesting to describe a scope of systems, for which Eidelmann's statement holds true.

Thus, let us consider problems of the second and third levels referring to the problem of interest to us (2.2.6), (2.2.7′):

$$\dot{y} = A(p,\xi)y(x), \quad x > 0, \qquad I^{(bN)}y(0) = \varphi; \qquad (2.2.11')$$

$$\mathcal{F} = M\tilde{\Phi}(0) = \tilde{\Phi}(0) = [Z(0), Z'(0), \dots, Z^{(r-1)}(0)]^{\top}. \qquad (2.2.14')$$

Here, $\det \mathcal{F} = W(0,p,\xi)$ coincides with the values at zero of Wronskian for a set of bounded solutions (2.2.10). Therefore, it differs from zero in x and it remains to be found out whether it has $p = p(\xi)$ roots from the range (2.2.9).

2. Working formula for completing condition ($N = 2$, $b = 1$). A. If $\sigma = 0(\tau_1^+ = \tau_2^+ = i\lambda)$, then two independent bounded solutions of equation (2.1.9) can be written as

$$u_1(x) = u_1(p,x,\xi) = \mathrm{e}^{-\lambda x},$$

$$u_2(p,x,\xi) = x\mathrm{e}^{-\lambda x}, \qquad \lambda = \sqrt{p+\xi^2}. \qquad (2.3.1)$$

By using formula (2.1.11) we determine functions $v_1(x)$ and $v_2(x)$ corresponding to them and compose the matrix $Z(x)$. Then

$$\det \mathcal{F} = \det Z(0) = \begin{vmatrix} 1 & 0 \\ v_1(0) & v_2(0) \end{vmatrix} = v_2(0). \qquad (2.3.2)$$

B. When $\sigma \neq 0$ we have

$$u_1(x) = \mathrm{e}^{\lambda^+ x}, \qquad u_2(x) = \mathrm{e}^{\lambda^- x}, \tag{2.3.3}$$

where $\lambda^\pm = -\sqrt{\xi^2 + p(\alpha/(2\beta) \pm \sigma)}$, $\operatorname{Re}\lambda^\pm < 0$.

The components $v_1(x)$ and $v_2(x)$ can be determined using (2.1.11):

$$\det \mathcal{F} = \begin{vmatrix} 1 & 1 \\ v_1(0) & v_2(0) \end{vmatrix} = v_2(0) - v_1(0). \tag{2.3.4}$$

3. Refinement of Eidelmann's hypothesis.

Lemma 2.3.1. *Eidelmann's hypothesis holds true when there is no mixed derivative in parabolic system in Petrovskii's sense.*

Proof. Let us prove the lemma for the second-order system ($N = 2$, $b = 1$) and for one higher-order equation ($N = 2$, $b \geq 2$).

Write a system of two equations with real coefficients of the form (2.1.8)

$$u_t = au_{xx} + \sum_{j=1}^{n} a_j \partial^2_{by} u + bv_{xx} + \sum_{j=1}^{n} b_j \partial^2_{y_j} v,$$
$$v_t = cu_{xx} + \sum_{j=1}^{n} c_j \partial^2_{y_j} u + dv_{xx} + \sum_{j=1}^{n} d_j \partial^2_{y_j} v.$$

To it there corresponds an ODE with respect to x

$$p\tilde{u} = a\tilde{u}'' - (\boldsymbol{a}, \xi^2)\tilde{u} + b\tilde{v}'' - (\boldsymbol{b}, \xi^2)\tilde{v},$$
$$p\tilde{v} = c\tilde{u}'' - (\boldsymbol{c}, \xi^2)\tilde{u} + d\tilde{v}'' - (\boldsymbol{d}, \xi^2)\tilde{v},$$

where the scalar product

$$(\boldsymbol{d}, \xi^2) = \sum_{j=1}^{n} d_j \xi_i^2.$$

Since $\beta = ab - cd > 0$, this system can be reduced to the canonical form, whence we can obtain the equality for the definition $\tilde{v}$

$$[\tilde{b}p + (\tilde{d}, \xi^2)]\tilde{v} = b\tilde{u}\text{"} - [\tilde{a}p + (\tilde{c}, \xi^2)]\tilde{u}.$$

Therefore, if $\sigma = 0$, then (2.3.2) $\tilde{v}_2(0) = \beta\lambda^2 \neq 0$ and if $\sigma \neq 0$, then

$$\tilde{v}_2(0) - \tilde{v}_1(0) \sim \lambda_+^2 + \lambda_-^2 \neq 0. \tag{2.3.5}$$

Lemma has been proved for the system.

Let us now consider a higher-order parabolic equation of the form

$$u_t = (-1)^{b-1}\Big(\partial_x^{2b} u + \sum_{j=1}^{n} \partial_{y_j}^{2b} u\Big), \qquad b \geq 2.$$

The characteristic equation $p + \xi^{2b} + \tau^{2b} = 0$ has precisely $2b$ various roots calculated by the formula

$$\tau_k = \sqrt[2b]{|p+\xi^{2b}|}\Big(\cos\frac{\varphi + 2k\pi}{2b} + i\sin\frac{\varphi+2k\pi}{2b}\Big) = q\psi_k \quad (\psi_i \neq \psi_j,\ i \neq j),$$

where $\varphi = \arg\left[-(p+\xi^{2b})\right]$, $k = 0, 1, \ldots, 2b-1$, module $q = |p+\xi^{2b}|^{1/(2b)} \neq 0$.

The decreasing term of the ODE fundamental solutions

$$(-1)^{b-1}\partial_x^{2b}\tilde{u} = (p+\xi^{2b})\tilde{u}$$

consists of the functions of the form

$$\tilde{u}_m(x,p,\xi) = \{e^{\lambda_m x}\} = \{e^{i\tau_m^+ x}\}, \quad m = 1, \ldots, b, \\ \tau_i^+ \neq \tau_j^+, \qquad i \neq j, \tag{2.2.10$'$}$$

to which there corresponds matrix (2.2.13)

$$\tilde{\Phi}(x) = \{\partial_x^k \tilde{u}_m\}, \qquad k = 0, 1, \ldots, b-1, \quad m = 1, \ldots, b.$$

Its determinant (Wronskian) can readily be calculated

$$W(x,p,\xi) = \exp\Big(\sum_{j=1}^{b} \lambda_j x\Big) \prod_{1\leq s<j<b} (\lambda_j - \lambda_s).$$

Consequently,

$$\det\mathcal{F} = W(0,p,\xi) = \prod_{1\leq s<j<b}(\lambda_j - \lambda_s) = q^b \prod_{1\leq s<j<b}(\psi_j - \psi_s) \\ = \sqrt{|p+\xi^{2b}|}\prod_{1\leq s<j<b}(\psi_j-\psi_s) \neq 0.$$

The lemma is proved. □

4. Illustration example of Lemma 2.3.1.

1. For Solonnikov's example (2.1.5.1a) formula (2.1.11) will be as follows:

$$4a^2\xi^2 v = -u" - 2a\xi u' + (p - a^2\xi^2 - \mu\xi^2)u.$$

A simple calculation according to (2.3.1), (2.3.2) yields

$$\det \mathcal{F} = v_2(0) = (\sqrt{p+\xi^2} - a\xi)/(2a^2\xi^2).$$

Therefore, $\det \mathcal{F} = 0$, provided

$$\sqrt{p+\xi^2} = a\xi,\ \xi > 0 \quad \text{or} \quad p = \mu\xi^2,\ \xi > 0. \tag{2.3.6}$$

Equality (2.3.6) coincides with the condition of Solonnikov (1965), under which the completing condition is violated. It is easily seen that the corresponding homogeneous problem

$$u'' + 2a\xi u' + a^2\xi^2 u = 0, \qquad u(0) = 0$$

admits a nontrivial solution $(u, v) = (x\mathrm{e}^{-a\xi x}, 0)$, $\xi > 0$.

2. For system (2.1.5.1b), we have

$$4a^2\xi^2 v = -u'' + (p - a^2\xi^2 - \mu\xi^2)u.$$

Here the completing condition holds true because

$$\det \mathcal{F} = v_2(0) = \lambda/(2a^2\xi^2) \neq 0$$

The homogeneous problem for it

$$pu - u'' - (a^2 + \mu)\xi^2 u = 0, \quad u'' + a^2\xi^2 u = 0, \quad u(0) = 0$$

admits only trivial solution $(u, v) = (0, 0)$.

3. For the system (2.1.52a) formula (2.1.11) takes the form

$$[(a_1 - a_3)\xi^2 + p]v = (\beta\lambda^2 + \gamma\beta\xi\lambda - \gamma\beta\xi^2)\mathrm{e}^{\lambda x}, \qquad \gamma = a_5/a_4 \neq 0.$$

It can easily be calculated that $\det \mathcal{F} = 0$, if $\lambda^+ + \lambda^- + \gamma\xi = 0$.

Let us study the resulting equality. Introducing

$$z = \sqrt{\xi^2 + \alpha p/(2\beta) - \sigma p} \equiv -\lambda^-$$

we obtain $\gamma\xi - z = \sqrt{z^2 + 2p\sigma}$. Squaring the both sides of the equality we obtain

$$\gamma^2\xi^2 - 2p\sigma = 2\gamma\xi z.$$

One more squaring gives us the equation

$$4\sigma^2 p^2 - 2\gamma^2\xi^2\alpha p/\beta + \gamma^2\xi^4(\gamma^2 - 4) = 0.$$

Hence we find

$$4\sigma^2(p)_{1,2} = \xi^2(\alpha\gamma^2/\beta \pm 2\gamma\sqrt{D}), \qquad D = 4\sigma^2 + \gamma^2/\beta > 0. \tag{2.3.7}$$

Choosing "+" or "−" in accordance with the sign of γ we can obtain that $p > 0$, which implies that the completing condition is violated.

4. Finally, for the system (2.1.5.2b) we obtain

$$[(a_1 - a_3)\xi^2 + p]v = -(b_1\lambda^2 + b_3\xi^2)e^{\lambda x}.$$

It implies $\det \mathcal{F} \neq 0$, because its value $\sim (\lambda^+)^2 + (\lambda^-)^2 \neq 0$. Therefore, the completing condition is fulfilled.

The performed analysis of the examples is in perfect agreement with Lemma 2.3.1.

2.3.2. Other types of parabolic systems

1. The first boundary-value problem can also be posed for general parabolic system provided $r = N$, for example, for parabolic system in Shirota's sense. Consider the system we already know (see (2.1.5.1b)):

$$\begin{aligned} u_t &= a_1 u_{xx} - i a_2 v_{xxx}, \\ v_t &= a_3 u_x + a_4 v_{xx}, \end{aligned} \tag{2.3.8}$$

for which

$$\alpha = a_1 + a_4 > 0, \qquad \beta = a_1 a_4 - a_2 a_3 > 0, \tag{2.3.9}$$

roots $\tau_{1,2}^+ = i\sqrt{p}\sqrt{\alpha/(2\beta) \pm \sigma}$ are different if $\sigma \neq 0$. We can obtain a higher-order equation (2.1.9), but then from the parameterized system

$$\begin{aligned} p\bar{u} - a_1\bar{u}'' + i a_2 \bar{v}_{xxx} &= 0, \\ p\bar{v} - i a_3 \bar{u}' &= a_4 \bar{v}'', \end{aligned} \tag{2.3.8'}$$

it will not be possible to find $\bar{v}$. We can only determine its derivative

$$a_2\bar{v}' = i a_4 \bar{u} - (i\beta/p) a_4 \bar{u}''.$$

The second disadvantage of system (2.3.8′) is that it cannot be canonized, i. e. it is unsolvable with respect to higher derivatives. Calculation shows that

$$\det Z(0) = i\sqrt{\beta}(a_4 + \sqrt{\beta})(\lambda^- - \lambda^+)/(pa_2).$$

Therefore, $\det Z(0) = 0$, provided that

$$a_4 = -\sqrt{\beta}. \tag{2.3.10}$$

Equality (2.3.10) is at variance with the condition of Solonnikov (1965) only because there was a misprint in this publication. The parabolicity preservation is also verified following the procedure given in this paper. Using the inequality $a_1 + a_4 > 0$ we obtain $\sqrt{\beta} < a_1$ and $a_1 > 0$. Taking into account (2.3.9), (2.3.10) we obtain $\beta + a_1\sqrt{\beta} + a_2a_3 = 0$. It is possible if the quadratic equation $x^2 + a_1x + a_2a_3 = 0$ has a root in the range $(0, a_1)$. We find the root $x_0 = -a_1/2 + \sqrt{a_1^2/4 - a_2a_3}$ and it will be the sought one if there hold the inequalities $0 < a_1/2 < \sqrt{a_1^2/4 - a_2a_3} < 3a_1/2$. By squaring them we obtain simple estimates

$$-2a_1^2 < a_2a_3 < 0.$$

Thus, if the following relations are true:

$$a_4 = -\sqrt{\beta} \qquad \text{or}$$

$$(a_1 > 0 \cap -2a_1^2 < a_2a_3 < 0 \cap a_4 = a_1/2 - \sqrt{a_1^2/4 - a_2a_3}\,) \tag{2.3.10′}$$

then the competing is violated even with respect to x.

2. The system of equations (see Subsection 2.1.2)

$$\begin{aligned} a_1u + a_2v_t &= a_3\Delta v, \\ a_3u_t + a_4(v_{tt} + \Delta^2 v) &= a_2\Delta u \end{aligned} \tag{2.3.11}$$

is parabolic in Solonnikov's sense if

$$\alpha = a_2^2 + a_3^2 > 0, \qquad \beta = a_2a_3 - a_1a_4 > 0$$
$$(N = r = 1, \quad s_1 = -2, \quad s_2 = 0, \quad t_1 = 2, \quad t_2 = 4).$$

The characteristic equation of the system has simple $\tau_{1,2}^+$-roots, when $\sigma = \alpha^2 - 4\beta^2 > 0$. Again, the respective (2.3.11), parameterized ODE are unsolvable with respect to higher derivatives, since those derivatives are present only in the second equation. Therefore, the following artificial multistep method is suggested:

1. The first equation is acted by the operator $a_4\Delta$ ($a_4 \neq 0$):

$$a_3a_4\Delta^2 v = a_2a_4\Delta v_t + a_1a_4\Delta v,$$

and the second equation is multiplied by $a_3 \neq 0$:

$$a_3a_4\Delta^2 v = -a_3^2 u_t + a_2a_3\Delta u - a_3a_4v_{tt}.$$

By comparing the resulting equations we obtain

$$a_2a_4\Delta v_t = \beta\Delta u - a_3^2 u_t - a_3a_4v_{tt}.$$

2. The first equation of (2.3.11) is acted by the operator $a_2a_4\partial_t$

$$a_2^2a_4v_{tt} + a_1a_2a_4u_t = a_2a_3a_4\Delta v_t,$$

hence by virtue of the equality from step 1 we obtain

$$a_3\beta\Delta u = a_4\alpha v_{tt} + (a_1a_2a_4 + a_3^2)v_t.$$

By adding to it the first equation of (2.3.11) we can write a new system

$$\begin{aligned} a_3\beta\Delta u &= a_4\alpha v_{tt} + (a_1a_2a_4 + a_3^2)v_t, \\ a_3\Delta v &= a_2v_t + a_1u. \end{aligned} \tag{2.3.11$'$}$$

3. Parameterize system (2.3.11′)

$$\begin{aligned} a_3\beta(\tilde{u}_{xx} - \xi^2\tilde{u}) &= a_4\alpha p^2\tilde{v} + (a_3\alpha - a_2\beta)p\tilde{u}, \\ a_3(\tilde{v}_{xx} - \xi^2\tilde{v}) &= a_2p\tilde{v} + a_1\tilde{u}. \end{aligned}$$

4. Rewrite the latter in the canonical form

$$\begin{aligned} \tilde{u}'' &= \Big(\frac{\xi^2}{p} - \frac{a_2}{a_3} + \frac{\alpha}{\beta}\Big)p\tilde{u} + \frac{a_4}{a_3}\frac{\alpha}{\beta}p^2\tilde{v}, \\ \tilde{v}'' &= \frac{a_1}{a_3}\tilde{u} + \Big(\xi^2 + \frac{a_2}{a_3}p\Big)\tilde{v}. \end{aligned}$$

5. Reduce the system to a single equation with respect to $\tilde{u}$

$$\beta\tilde{u}^{(\mathrm{IV})} - (\alpha p + 2\beta\xi^2)\tilde{u}'' + \beta(p^2 + \alpha p\xi^2/\beta + \xi^4)\tilde{u} = 0$$

and to the equality with respect to $\tilde{v}$

$$\tilde{v} = \frac{a_3}{\alpha a_4 p^2}\Big[\beta\tilde{u}'' - p\Big(a_3^3 + \frac{a_1 a_2 a_4}{a_3} + \beta\frac{\xi^2}{p}\Big)\tilde{u}\Big].$$

Performing simple calculation we arrive at

$$\det Z(0) = \frac{a_3}{\alpha a_4 p}(\lambda_-^2 - \lambda_+^2) \neq 0,$$

where $a_3 \neq 0$, $a_4 \neq 0$.

The value of the determinant with respect to both x and p is nonzero, therefore the completing condition is satisfied.

3. Calculating the limiting value of integrals. Let us consider integrals of the potential type for $x = 0$:

$$\begin{aligned} W_a^{(k,m)}(t,0,y) &\\ = \lim_{x\to 0} x^m \int_0^t \mathrm{d}\tau \int_{\mathbb{R}^q} f(\tau,z)\exp\Big(-\frac{x^2+(y-z)^2}{4a(t-\tau)}\Big)\frac{\mathrm{d}z}{(y-z)^k(t-\tau)^s} &\\ = \lim_{x\to 0} x^m \int_0^t \int_{\mathbb{R}^q} f(\tau,z)K(t-\tau,x,y-z)\,\mathrm{d}z\,\mathrm{d}\tau, \end{aligned}$$

where the following values of parameters are admissible:

$$m \geq 0, \qquad 0 \leq k_j < 1, \qquad s > 0, \qquad q \geq 0.$$

When $q \geq 1$ changing of the variable z_j by ξ_j ($j = 1,\dots,q$) by the formula

$$z_j = y_j - 2\xi_j\sqrt{a(t-\tau)}, \qquad \mathrm{d}z_j = -2\sqrt{a(t-\tau)}\,\mathrm{d}\xi_j$$

we obtain

$$\begin{aligned} W_a^{(k,m)} = (4a)^{(q-k)/2}\lim_{x\to 0} x^m \int_{\mathbb{R}^q}\frac{\mathrm{e}^{-\xi^2}}{\xi^k}\,\mathrm{d}\xi \int_0^t f(\tau, y - 2\sqrt{a(t-\tau)}\,\xi) &\\ \times\exp\Big(-\frac{x^2}{4a(t-\tau)}\Big)\frac{\mathrm{d}\tau}{(t-\tau)^{s+(k-q)/2}}. \end{aligned}$$

Further, substitution $\tau \to \eta$,

$$x = 2\eta\sqrt{a(t-\tau)}, \quad \mathrm{d}\tau = \frac{x^2}{2a\eta^3}\,\mathrm{d}\eta, \quad (t-\tau)^{-\nu/2} = \Big(\frac{2\eta\sqrt{a}}{x}\Big)^{\nu}$$

yields

$$W_a^{(k,m)} = 2^{2s-q}a^{s-q}\lim_{x\to 0} x^{\sigma}\int_{\mathbb{R}^q}\frac{\mathrm{e}^{-\xi^2}}{\xi^k}\,\mathrm{d}\xi\int_{x/(2\sqrt{at})}^{\infty} f\Big(t-\frac{x^2}{4a\eta^2}, y-\frac{x\xi}{\eta}\Big) \times \eta^{2s+k-4}\mathrm{e}^{-\eta^2}\,\mathrm{d}\eta,$$

where $\sigma = m+q+2-k-2s$. For convergence of the integrals it is required that

$$0 \le k_j < 1, \quad j=1,\dots,q; \qquad 2s+k>3.$$

Lemma 2.3.2. *The expression for the limiting value of the integrals is*

$$W_a^{(k,m)} = \begin{cases} \infty, & \sigma<0, \\ \mathrm{const}\cdot f(t,y), & \sigma=0, \\ 0, & \sigma>0, \end{cases}$$

where

$$\mathrm{const} = 2^{2s-q}a^{s-q}\int_0^{\infty}\eta^{2s+k-4}\mathrm{e}^{-\eta^2}\,\mathrm{d}\eta\int_{\mathbb{R}^q}\frac{\mathrm{e}^{-\xi^2}}{\xi^k}\,\mathrm{d}\xi \neq 0.$$

In order to better memorize the result, let us conventionally give these values such exotic names as "cosmos," "earth',' and "hole," respectively. Clearly, only the case of "earth" can be used for the calculation, i. e. when the finite jump takes place with the factor const $\neq 0$; in the "cosmos" version the integrals diverge, and in the case of "hole' we lose the information from the boundary data.

Therefore, the last two values will be brought to the "earth" level using the following procedure. In the case of "cosmos" we will first represent the kernel of the integral as

$$K(t-\tau, x, y-z) = F(\partial_\tau, \partial_z)K_1(t-\tau, x, y-z),$$

whereas to the new kernel K_1 the number $\sigma_1 = 0$ should correspond. Then by the partial integration the differential operator $F(\partial_\tau, \partial_z)$ will be transferred to the function $f(\tau, z)$.

For the case of "hole" using the uniform convergence of multidimensional integrals the substitution in the exponent $x=0$ must be performed when it is convenient, or, as in the "cosmos" case, the kernel should be transformed into a different (required) form.

Remark 2.3.1. In the both transformation variants fraction-order operators can arise.

4. Fraction-order operators. Fraction-order operators are a natural Liouville generalization to the functions of many variables of fractional differentiation D_t^α and fractional integration I^α (Siegmund, 1965). By way of example, let us take the equation of thermal conductivity

$$F_a u(t,x,y) \equiv [\partial_t - a(\partial_x^2 + \Delta_y)]u(t,x,y) = 0,$$
$$t > 0, \qquad x > 0, \qquad y \in \mathbb{R}^n, \quad n \geq 1.$$

Recall its fundamental solution

$$E_a(t,x,y) = \frac{1}{(4a\pi t)^{(n+1)/2}} \exp\Big(-\frac{x^2 + |y^2|}{4at}\Big).$$

Now let us write the related equation operators of fractional differentiation

$$F_a^\alpha f(t,y) = \frac{1}{\Gamma(1-\alpha)} F_a \int_0^t \mathrm{d}\tau \int_{\mathbb{R}^n} f(\tau,z)\, \frac{E_a(t-\tau, y-z)}{(t-\tau)^\alpha}\, \mathrm{d}z,$$

and fractional integration

$$F_a^{-\alpha} f(t,y) = \frac{1}{\Gamma(\alpha)} \int_0^t \mathrm{d}\tau \int_{\mathbb{R}^n} f(\tau,z)\, \frac{E_a(t-\tau, y-z)}{(t-\tau)^{1-\alpha}}\, \mathrm{d}z,$$

where $0 \leq \alpha \leq 1$, $\Gamma(\alpha)$ is the Gauss gamma-function.

Let us introduce thermal potentials:

A) The volume masses potential

$$F_a^{-1} f(t,y) = \int_0^t \mathrm{d}\tau \int_{\mathbb{R}^n} f(\tau,z) E_a(t-\tau, y-z)\, \mathrm{d}z$$

is characterized by the property

$$F_a(F_a^{-1} f) = f(t,y), \qquad \lim_{t \to +0} F_a^{-1} f = 0,$$

i. e., satisfies the heterogeneous equation and homogeneous Cauchy condition

$$F_a u = f, \qquad u(0,y) = 0.$$

Therefore, the generalization of the volume masses potential is the fractional operator $F_a^{-\alpha} f$.

B) The surface ($t = 0$) potential (Poisson integral)

$$F_a^{-1}(t, u_0(y)) = \int_{\mathbb{R}^n} u_0(z) E_a(t, y - z)\, \mathrm{d}z$$

solves the Cauchy problem $F_a^u = 0$, $u(0, y) = u_0(y)$. The generalization of Poisson integral operator is introduced in the form of iterated integration

$$F_a^{-m}(t, u_0(y)) = F_a^{-1}[F_a^{(1-m)}(t, u_0(y))].$$

For iteration of the operator F_a^{-1} the following formula holds true:

$$F_a^{-m}(t, u_0(y)) = F_{ma}^{-1}(t, u_0(y)).$$

2.3.3. Solution of conditionally well-posed problems

1. Problem for parabolic system in Shirota's sense (2.3.8) with "coefficient" condition of ill-posedness

$$H = (a_4 = -\sqrt{\beta}). \tag{2.3.9'}$$

A) Normal solution. Step 1. SLAE solution. Let us write a SLAE

$$\begin{aligned} C_1 + C_2 &= \bar{f}_1(p), \\ i(M^+C_1 + M^-C_2)/b &= \bar{f}_2(p), \end{aligned} \tag{2.3.11.1}$$

where

$$M^\pm = \sqrt{\beta/p}\, A^\pm, \qquad A^\pm = a_\pm^{-1/2} + (\beta a_\pm)^{1/2},$$
$$a_\pm = \alpha/(2\beta) \pm \sigma, \qquad \sigma = \sqrt{\alpha^2/(4\beta^2) - 4\beta} > 0.$$

By the condition H (2.3.9$'$) the principal determinant of the system $\Delta = i(M^- - M^+)/a_2 = 0$, therefore let us require that the other determinants of the system's extended matrix should also become zero ($\Delta_1 = \Delta_2 = 0$). This leads to the condition of solvability

$$\bar{f}_2(p) = (iM^+/a_2)\bar{f}_1(p). \tag{2.3.12.1}$$

At the same time system (2.3.11.1) degenerates into a single equation. Let us find its normal solution:

$$I(C_2) = \left\| \begin{matrix} \bar{f}_1 - C_2 \\ C_2 \end{matrix} \right\|^2 \quad \Rightarrow \quad \inf_{C_2} : C_2^n = \bar{f}_1,\ C_1^n = 0,$$

and $\inf I(C_2^n) = \bar{f}_1^2$.

It is clear that for such C^n, equalities (2.3.11.1) are completely satisfied due to the solvability condition (2.3.12.1).

Step 2. Let us define the vector corresponding to the obtained "constant"

$$(\bar{u}_n, \bar{v}_n) = Z(x)C^n = \begin{bmatrix} e^{-x\sqrt{pa_+}} & e^{-x\sqrt{pa_-}} \\ iM^+ e^{-x\sqrt{pa_+}}/a_2 & iM^- e^{-x\sqrt{pa_-}}/a_2 \end{bmatrix} \begin{bmatrix} 0 \\ \bar{f}_1 \end{bmatrix}$$
$$= (\bar{f}_1, \bar{f}_2) e^{-x\sqrt{pa_-}} \qquad (2.3.13)$$

which is exact solution of the first boundary-value problem for (2.3.8). The statement is confirmed by the substitution of (2.3.13) in the system with allowance for (2.3.12.1).

Step 3. Calculation of originals. Using the Borel theorem of convolution of two functions and formula (2.1.1) of Bateman and Erdélyi (1969), we obtain

$$u_n(t,x) = \int_0^t f_1(\tau) G(t-\tau, x)\, d\tau,$$

where

$$G(t,x) = \Lambda_p^-(e^{-x\sqrt{pa_-}}) = \frac{x}{t}\sqrt{\frac{a_-}{4\pi t}}\, e^{-a_- x^2/(4t)}.$$

Recall the assumed denotation for the fundamental solution of thermal conductivity equation

$$E_{a_-}(t,x) = \frac{1}{2}\sqrt{\frac{a_-}{\pi t}}\, e^{-a_- x^2/(4t)}.$$

We have,

$$u_n(t,x) = R(f_1), \qquad v_n = R(f_2). \qquad (2.3.14)$$

Here

$$R(f_j) = -\partial_x \int_0^t f_j(\tau) E_{a_-}(t-\tau, x)\, d\tau, \qquad j = 1, 2.$$

The condition (2.3.12) rewritten in terms of originals takes the form

$$f_2(t) = \frac{i\sqrt{\beta}A^+}{a_2} \int_0^t \frac{f_1(\tau)}{\sqrt{\pi(t-\tau)}}\, d\tau \equiv \frac{i\sqrt{\beta}A^+}{a_2} I^{1/2}(f_1). \qquad (2.3.12.1')$$

Hence we obtain the relation for the function $f_1(t)$

$$\frac{i\sqrt{\beta}A^+}{a_2} f_1(t) = \partial_t \int_0^t \frac{f_2(\tau)}{\sqrt{\pi(t-\tau)}}\, d\tau \equiv D^{1/2}(f_2). \qquad (2.3.12.1'')$$

Here $I^{1/2}$ and $D_t^{1/2}$ are integral and differential operators of the order 1/2 respectively.

Let us prove that the vector (2.3.14) is the classical solution of the first boundary-value problem for system (2.3.15). It can be seen from (2.3.14) that $\partial_t R(f) = \partial_x^2 R(f)/a_-$, therefore (2.3.8) takes the form

$$\begin{aligned} u_t &= a_1 a_- u_t - i a_2 a_- v_{tx}, \\ (1/a_- - a_4) v_{xx} &= i a_3 v_x, \end{aligned}$$

or

$$\begin{aligned} (1 - a_1 a_-) u + i a_2 a_- v_x &= 0, \\ i a_3 a_- u + (a_4 a_- - 1) v_x &= 0. \end{aligned}$$

Since the determinant of the latter one equal to zero

$$\Delta = (1 - a_1 a_-)(1 - a_4 a_-) - a_2 a_3 a_-^2 = 0,$$

it is sufficient to study only one equation

$$(1 - a_1 a_-) u = -i a_2 a_- v_x.$$

According to the theory of parabolic operators $v_x = (a_- \partial_t)^{1/2} v$, therefore

$$(1 - a_1 a_-) f_1(\tau) - i a_2 a_- (a_- \partial_t)^{1/2} f_2(\tau) = [(1 - a_1 a_-) - a_- \sqrt{\beta a_-} A^+] f_1 = 0.$$

In calculation we use (2.3.12.1″) and equality of the values of A^+ and a_-.

On the boundary ($x = 0$) we have

$$\frac{\sqrt{a_-}}{2\sqrt{\pi}} \lim_{x \to 0} x \int_0^t f(\tau) \frac{e^{-x^2 a_-/4(t-\tau)}}{(t-\tau)^{3/2}} \, d\tau = \left| \eta = \frac{x}{2} \sqrt{\frac{a_-}{t-\tau}} \right| = f(t),$$

which proves the claim.

B) Pseudosolution. If the functions $f_1(t)$ and $f_2(t)$ are retained arbitrary, then it is necessary to find a pseudosolution of SLAE, which follows from the variational problem

$$I(C) = \|\mathcal{F} c - f\| \quad \Rightarrow \quad \min_C .$$

The condition of the extremum $\partial_{C_j} I(C) = 0$ ($j = 1, 2$) is equivalent to one equation

$$\Big(1 - \Big(\frac{M^+}{a_2}\Big)^2\Big) C_1 + \Big(1 - \frac{M^+ M^-}{a_2^2}\Big) C_2 = \bar{f}_1 + \frac{i M^+}{a_2} \bar{f}_2$$

Hence we find the set of pseudosolutions $C_1^p = D - dC_2$, where the following notation is introduced:

$$d = \frac{a_2^2 - M^+M^-}{a_2^2 - M^{+2}}, \qquad D = \frac{a_2^2\bar{f}_1 + ia_2M^+\bar{f}_2}{a_2^2 - M^{+2}}.$$

To provide the uniqueness, we define the normal pseudosolution

$$I(C_2) = \left\| \begin{matrix} D - dC_2 \\ C_2 \end{matrix} \right\|^2 \Rightarrow \inf_{C_2}: \quad C_1^n = \frac{D}{1+d^2}, \quad C_2^n = \frac{2D}{1+d^2}$$

where $\inf I(C_2^n) = D^2/(1+d^2)$.

Let us write out the obtained values of the coefficients

$$C_j^n(p) = \frac{a_2 d_j}{d_1^2 + d_2^2}(a_2\bar{f}_1(p) + iM^+\bar{f}_2(p)), \qquad j = 1, 2$$

where $d_1^2 = a_2^2 - M^{+2}$, $d_2^2 = a_2^2 - M^+M^-$.

Step 2′. The solution of the boundary-value problem for the ODE is

$$\begin{pmatrix} \bar{u}_n \\ \bar{v}_n \end{pmatrix}(p, x)$$

$$= \left\{ \begin{matrix} a_2(d_1 \mathrm{e}^{-x\sqrt{pa_+}} + d_2 \mathrm{e}^{-x\sqrt{pa_-}}) \\ i(d_1M^+\mathrm{e}^{-x\sqrt{pa_+}} + d_2M^-\mathrm{e}^{-x\sqrt{pa_-}}) \end{matrix} \right\} \frac{a_2\bar{f}_1(p) + iM^+\bar{f}_2(p)}{d_1^2 + d_2^2}.$$

Step 3′. It remains to apply to the latter the inverse operator $\Lambda_p^-(x)$.

2. Solonnikov's counter-example. Consider the problem

$$\begin{aligned} u_t &= u_{xx} - 2iau_{xy} - (a^2 + \mu)u_{yy} - 4a^2v_{yy}, \\ v_t &= -u_{xx} + 2iau_{xy} + a^2u_{yy} + \Delta v + +2iav_{xy} + 2a^2v_{yy}, \\ u(t,0,y) &= f_1(t,y), \quad v(t,0,y) = f_2(t,y), \quad (t,y) \in \mathbb{R}^2, \\ a &= \sqrt{1+\mu}, \qquad \mu > 0 \end{aligned} \tag{2.3.15}$$

provided the "root" condition of ill-posedness

$$p = \mu\xi^2, \qquad \xi > 0, \quad \mu > 0. \tag{2.3.16}$$

i) To this problem there corresponds the following SLAE

$$\begin{aligned} C_1 &= \tilde{f}_1(p,\xi), \\ b(\lambda - b)C_1 + (\lambda - b)C_2 &= 2b^2\tilde{f}_2(p,\xi). \end{aligned} \tag{2.3.11.2}$$

Here $\lambda = \sqrt{p + \xi^2}$, $b = a\xi$.

Under solvability condition

$$\tilde{f}_2(\mu\xi^2, \xi) = 0, \tag{2.3.12.2}$$

obtained according to (2.3.16), the solution of the SLAE is $C_1 = \tilde{f}_1$, C_2 is arbitrary. Hence the normal solution is

$$C^n = (\tilde{f}_1, 0). \tag{2.2.1}$$

ii) Consider the problem for parametrized ODE

$$\begin{gathered}\tilde{u}'' + 2a\xi\tilde{u}' + (a^2 + \mu)\xi^2\tilde{u} - p\tilde{u} + 4a^2\xi^2\tilde{v} = 0,\\ \tilde{u}'' - \tilde{v}'' + 2a\xi(\tilde{u}' + \tilde{v}') + a^2\xi^2\tilde{u} + (p + \xi^2 + 2a^2\xi^2)\tilde{v} = 0,\\ \tilde{u}(0) = \tilde{f}_1, \qquad \tilde{v}(0) = \tilde{f}_2.\end{gathered} \tag{2.3.15$'$}$$

The solution is found by the formula

$$(\tilde{u}, \tilde{v})_n = \tilde{\Phi}(x)C^n = \mathrm{e}^{-\lambda x}\begin{pmatrix} 1 & x \\ \dfrac{\lambda - a\xi}{2a\xi} & \dfrac{\lambda - a\xi}{2a\xi}\left(x + \dfrac{1}{a\xi}\right)\end{pmatrix}\begin{pmatrix}\tilde{f}_1\\ 0\end{pmatrix}$$
$$= \left(1, \frac{\lambda - a\xi}{2a\xi}\right)\mathrm{e}^{-\lambda x}\tilde{f}_1.$$

It satisfies the system of ODE (2.3.15$'$) and the following equality at $x = 0$

$$\tilde{u}_n(p, 0, \xi) = \tilde{f}_1(p, \xi), \qquad \tilde{v}_n(p, 0, \xi) = \frac{\lambda - a\xi}{2a\xi}\,\tilde{f}_1(p, \xi).$$

The boundary condition for $\tilde{u}_n$ holds everywhere, and for $\tilde{v}_n$ it holds only along the right branch of parabola (2.3.16), when there holds (2.3.12.2).

iii) The solution of the original problem.

A. Transformation of (2.3.12.2) proceeds in a nonstandard way

$$\hat{f}_2(p, \xi)\big|_{p=\mu\xi^2} \overset{\Phi_\xi^-}{=} \bar{f}_2(\mu\xi^2, y) \overset{\Delta}{=} \int_{\mathbb{R}} f_2(t, y)\mathrm{e}^{-\mu\xi^2 t}\,\mathrm{d}t = \int_0^\infty f_2(t, y)\mathrm{e}^{-\mu\xi^2 t}\,\mathrm{d}t,$$

since according to the theory of two-sided Laplace transform ($t \in R$) it is required $f_2(t, y) = 0$ for $t < 0$. Therefore, (2.3.12.2) implies

$$\int_0^\infty f_2(t, y)\mathrm{e}^{-st}\,\mathrm{d}t = 0, \qquad s > 0, \quad y \in \mathbb{R}.$$

Consequently

$$f_2(t,y) = 0. \tag{2.3.12.2}$$

B. Let us calculate the inverse transforms

$$\hat{u}_n = \mathrm{e}^{-\lambda x}\hat{f}_1(p,\xi), \qquad \hat{v}_n = -\frac{1}{2}\Big(1+\frac{1}{a\xi}\,\partial_x\Big)\hat{u}_n, \qquad \lambda = \sqrt{p+\xi^2}.$$

Taking into account of the well-known formula for convolution of the integral transform of two functions, we obtain

$$\begin{aligned} u_n(t,x,y) &= \int_0^t \mathrm{d}\tau \int_{\mathbb{R}} f_1(\tau,z)G(t-\tau,x,y-z)\,\mathrm{d}z \\ &= -2\partial_x \int_0^t \mathrm{d}\tau \int_{\mathbb{R}} f_1(\tau,z)E(t-\tau,x,y-z)\,\mathrm{d}z, \end{aligned} \tag{2.3.17}$$

where the kernel

$$G(t-\tau,x,y-z) = \Lambda_p^-(t)\Phi_\xi^-(y)[\mathrm{e}^{-\lambda x}] = -2\partial_x E(t-\tau,x,y-z).$$

Let us proof (2.3.17). Indeed,

$$\Phi_\xi^-(\mathrm{e}^{-\lambda x}) = 2\int_0^\infty \mathrm{e}^{-\lambda x}\cos(y\xi)\,\mathrm{d}\xi = \frac{2x}{\sqrt{x^2+y^2}}\sqrt{p}K_1(\sqrt{p(x^2+y^2)}\,).$$

Prepare the last factor for the next stage

$$\sqrt{p}K_1(\sqrt{\alpha p}) = \frac{p}{\sqrt{\alpha}}\Big[\frac{\sqrt{\alpha}K_1(\sqrt{\alpha p})}{\sqrt{p}}\Big] \overset{\Lambda_p^-}{\Rightarrow} \partial_t \frac{\mathrm{e}^{-\alpha/(4t)}}{\sqrt{\alpha}},$$

where $\alpha = x^2+y^2$. Therefore,

$$u_n(t,x,y) = \frac{2}{\pi}\int_0^t \mathrm{d}\tau \int_{\mathbb{R}} f_1(\tau,z)K(t-\tau,x,y-z)\,\mathrm{d}z,$$

where

$$\begin{aligned} K(t,x,y) &= \frac{x}{x^2+y^2}\,\partial_t \mathrm{e}^{-(x^2+y^2)/(4t)} \\ &= \frac{x}{4t^2}\,\mathrm{e}^{-(x^2+y^2)/(4t)} = -\frac{1}{2t}\,\partial_x \mathrm{e}^{-(x^2+y^2)/(4t)}. \end{aligned}$$

Thus, formula (2.3.17) is proved.

Further, it will be additionally required to find the original of the fraction $\mathrm{e}^{-\lambda x}/\xi$. Again, according to the theorem on convolution we have

$$\Phi_\xi^-\Big[\frac{\mathrm{e}^{-\lambda x}}{\xi}\Big] = \Phi_\xi^-(\mathrm{e}^{-\lambda x}) * \Phi_\xi^-\Big(\frac{1}{\xi}\Big).$$

Let us calculate

$$\Phi_\xi^-\Big(\frac{1}{\xi}\Big) = -i\sqrt{\frac{2}{\pi}}\int_0^\infty \frac{\sin(\xi y)}{\xi}\,\mathrm{d}\xi = -\frac{i\pi}{2}.$$

Therefore,

$$\Phi_\xi^-\Big[\frac{\mathrm{e}^{-\lambda x}}{\xi}\Big] = \frac{i\pi x}{\sqrt{x^2+y^2}}\sqrt{p}K_1(\sqrt{p(x^2+y^2)})$$
$$= -i\pi x\int_0^\infty \frac{\sqrt{p}K_1(\sqrt{p(x^2+(y-\eta)^2)})}{\sqrt{x^2+(y-\eta)^2}}\,\mathrm{d}\eta.$$

Let us determine the auxiliary function

$$W(t,x,y) = \Lambda_p^-(t)\Phi_\xi^-\Big[\hat{f}_1(p,\xi)\frac{\mathrm{e}^{-\lambda x}}{\xi}\Big]$$
$$= -\frac{i\pi}{2}\int_0^t \mathrm{d}\tau\int_{\mathbb{R}} f_1(\tau,z)K_1(t-\tau,x,y-z)\,\mathrm{d}z,$$

where

$$K_1(t,x,y) = 2x\partial_t\int_0^\infty \exp\Big(-\frac{x^2+(y-\eta)^2}{4(t-\tau)}\Big)\frac{\mathrm{d}\eta}{x^2+(y-\eta)^2}$$
$$= -\frac{\partial_x}{t}\int_0^\infty \exp\Big(-\frac{x^2+(y-\eta)^2}{4t}\Big)\,\mathrm{d}\eta.$$

Finally, we get

$$W(t,x,y) = -\frac{i\pi\partial_x}{2}\int_0^\infty \mathrm{d}\eta$$
$$\times\Big\{\int_0^t \mathrm{d}\tau\int_{\mathbb{R}}\frac{f_1(\tau,z)}{t-\tau}\exp\Big(-\frac{x^2+(y-\eta-z)^2}{4(t-\tau)}\Big)\,\mathrm{d}z\Big\}$$
$$= 2i\int_0^\infty \mathrm{d}\eta\,\partial_x\Big\{\int_0^t \mathrm{d}\tau\int_{\mathbb{R}} f_1(\tau,z)E(t-\tau,x,y-\eta-z)\,\mathrm{d}z\Big\}.$$

Now we can write the component

$$v_n(t,x,y) = -\frac{u_n}{2} - \frac{1}{2a}F,$$

where $F = \partial_x W$.

C. Control of the "solution". Let us substitute v_n in the system of differential equations and taking into account that $\partial_t u_n = \Delta u_n$, $\partial_t F = \Delta F$ we get $F_{yy} = iu_{xy}$, $F_{xy} = iu_{xx}$. Obviously, they are true provided $F_y = iu_x$. Indeed

$$\begin{aligned} F_y &= W_{xy} = -W_{x\eta} \\ &= -2i\partial_x^2\Big(\int_0^t \mathrm{d}\tau \int_{\mathbb{R}} f_1(\tau,z)E_1(t-\tau,x,y-\eta-z)\big|_{\eta=0}^{\eta=\infty}\,\mathrm{d}z\Big) \\ &\qquad = 2i\partial_x^2\Big(\int_0^t \mathrm{d}\tau \int_{\mathbb{R}} f_1(\tau,z)E_1(t-\tau,x,y-z)\,\mathrm{d}z\Big) = iu_x. \end{aligned}$$

For $x=0$ by the lemma on limiting values of integrals, we get

$$u_n(t,0,y) = f_1(t,y), \qquad v_n(t,0,y) - \frac{f_1}{2} - \frac{1}{2a}F^0(t,y).$$

Further we determine

$$\begin{aligned} F^0 &= 2i\int_0^\infty \mathrm{d}\eta \int_0^t \mathrm{d}\tau \int_{\mathbb{R}} f_1(\tau,z)\partial_x^2 E_1(t-\tau,0,y-\eta-z)\,\mathrm{d}z \\ &= -2i\int_0^\infty \mathrm{d}\eta \int_0^t \mathrm{d}\tau \int_{\mathbb{R}} f_1(\tau,z)(\partial_\tau^+\partial_z^2)E_1(t-\tau,0,y-\eta-z)\,\mathrm{d}z \\ &= 2i\Big[-\int_0^t \mathrm{d}\tau \int_{\mathbb{R}} \partial_z f_1(\tau,z)E_1(t-\tau,0,y-\eta-z)\,\mathrm{d}z \\ &\qquad + \int_0^t \mathrm{d}\tau \int_{\mathbb{R}} \partial_\tau f_1(\tau,z)\Big(\int_0^\infty E_1(t-\tau,0,y-\eta-z)\,\mathrm{d}\eta\Big)\,\mathrm{d}z\Big], \end{aligned}$$

assuming that $f_1(o,z)=0$, in addition, we have

$$\int_0^\infty E_1(t-\tau,0,y-\eta-z)\,\mathrm{d}\eta = \frac{1}{4\sqrt{\pi(t-\tau)}}\Big[1-\Phi\Big(\frac{|y-z|}{2\sqrt{t-\tau}}\Big)\Big],$$

where $\Phi(x) = \mathrm{erf}\,(x)$.

Conclusion. When the "coefficient" condition is imposed, the problem has a unique classical solution (**1**), and when the "root" condition holds it has a "nonclassical" solution satisfying the system of differential equations and boundary data with a different right-hand side (**2**).

2.4. MIXED PROBLEM OF HEAT AND MASS EXCHANGE

2.4.1. Reducing the problem level

In the quadrant $D = (t > 0,\ x > 0,\ y \in \mathbb{R})$ consider the problem

$$\begin{gathered} u_t = a\Delta u, \qquad \nu_t = b\Delta v, \qquad 0 < a \leq b, \\ u(0,x,y) = v(0,x,y) = 0, \\ B(\partial_x)u\big|_{x=0} = f(t,y): \qquad \begin{cases} \alpha_0 u_x + \alpha_1 v_x = f_1(t,y), \\ \beta_0 u + \beta v = f_2(t,y), \end{cases} \end{gathered} \tag{2.4.1}$$

where the constants α_j, β_j satisfy the inequality (Basarbayeva, 1982)

$$\operatorname{rang}\begin{vmatrix} \alpha_0 & \alpha_1 & 0 & 0 \\ 0 & 0 & \beta_0 & \beta_1 \end{vmatrix} = 2.$$

Earlier, problem (2.4.1) had been studied by Basarbayeva (1982) and Temirbulatov (1989a,b). Particular, in Basarbayeva (1982) the "coefficient" condition is obtained and the unique solvability is proved in Holder classes when I-condition does not hold; it is noted that similar problems occur in the processes of heat and mass transfer in capillary-porous bodies under vacuum conditions. In Temirbulatov (1989a) by the modified method of heat potential uniqueness theorems is proved for an ill-posed variant in special sets. In Temirbulatov (1989b) a coefficient condition of ill-posedness is given, but the study was cut short at the stage of problem transform solution (SLAE is solved).

Further, integral transformations yield the problem for ODE in the variable x

$$\begin{gathered} \tilde{u}''(p,x,\xi) = (p/a + \xi^2)\tilde{u}, \\ \tilde{v}'' = (p/b + \xi^2)\tilde{v}(p,x,\xi), \qquad x > 0, \\ \alpha_0\tilde{u}' + \alpha_1\tilde{v}' = \tilde{f}_1(p,\xi),, \\ \beta_0\tilde{u} + \beta_1\tilde{v} = \tilde{f}_2(p,\xi), \qquad x = 0. \end{gathered} \tag{2.4.2}$$

From bounded linearly independent solutions of differential equation the matrix is composed

$$Z(x) = \begin{bmatrix} e^{-x\lambda_\alpha} & 0 \\ 0 & e^{-x\lambda_b} \end{bmatrix}, \qquad \lambda_a = \sqrt{\frac{p}{a} + \xi^2}.$$

By substituting the general solution of the equations

$$(\tilde{u}, \tilde{v}) = Z(x)C, \qquad C = (C_1, C_2) \tag{2.4.3}$$

into the boundary data, we arrive at the SLAE with respect to the vector C

$$\mathcal{F}C = \tilde{f}, \qquad \mathcal{F} = B(\partial)Z(x)\big|_{x=0} = \begin{bmatrix} -\alpha_0\lambda_a & -\alpha_0\lambda_b \\ \beta_0 & \beta_1 \end{bmatrix}. \tag{2.4.4}$$

2.4.2. Violation of the Lopatinskii condition

Let us consider the equality

$$\det \mathcal{F} = \alpha_1\beta_0\lambda_b - \alpha_0\beta_1\lambda_a = 0.$$

By introducing the quantity $\chi = \alpha_1\beta_0/\alpha_0\beta_1 \neq 0$, it can be simplified to $\lambda_a = \chi\lambda_b$. Squaring of the both sides of the latter we obtain

$$(1/a - \chi^2/b)p = (\chi^2 - 1)\xi^2, \tag{2.4.5}$$

which yields

$$p = \alpha\xi^2, \quad \alpha = a(\chi^2 - 1)/(1 - a\chi^2/b) \geq 0. \tag{2.4.6}$$

The following conclusions can be made:

1) The "domain" of ill-posedness is the segment $1 \leq \chi \leq \sqrt{b/a}$, since on it equality (2.4.5) holds.

2) "The set of zeros" of the determinant is as follows:

$$P_0 \equiv \begin{cases} \text{at the point } \chi = 1 \text{ equality (2.4.5) holds when } a = b, \\ (p \text{ and } \xi \text{ are arbitrary}); \text{ if } a \neq b, \text{ then } p = 0, \ \xi \text{ is} \\ \text{arbitrary;} \\ \text{in the interval } (1, \sqrt{b/a}) \text{ there holds (2.4.6);} \\ \text{at the point } \chi = \sqrt{b/a} \text{ we have } \xi = 0, \ p \text{ is arbitrary.} \end{cases}$$

3) The I-conditions of ill-posedness:

   α) "coefficient" condition: $(a = b) \cap (\chi = 1)$, p and ξ are arbitrary;

   β) "root" condition: parameters p and ξ are connected by (2.4.6);

   γ) "coefficient and root" condition: $\chi = \sqrt{b/a}$, $a \neq b$, the parameter $\xi = 0$, the parameter p is arbitrary.

4) The case of $p = 0$ and ξ is arbitrary, excluded from the consideration, because the value $p = 0$ does not belong to the range (2.2.9).

The relation from 1) coincides with the results obtained by Basarbayeva (1982), and the set of zeros P_0 coincides with the results obtained by Temirbulatov (1989a).

2.4.3. Ill-posed problems

Further, considering problems of various levels when the I-condition holds, those are denoted by letters α, β, or γ.

According to the definition of Tikhonov and Arsenin (1986) we obtain the normal solution of the SLAE (2.4.4):

$$C^n(p,\xi) = (C_1^n, C_2^n)$$
$$= \left(\frac{\beta_0}{\Delta}, \frac{\beta_1}{\Delta}\right) \times \begin{cases} F(p,\xi) = \tilde{f}_2(p,\xi), & (\alpha) \\ F(\xi) = \tilde{f}_2(\alpha\xi^2,\xi), & (\beta) \\ F(p) = \tilde{f}_2(p,0), \quad \Delta = \beta_0^2 + \beta_1^2 \neq 0, & (\gamma) \end{cases} \qquad (2.4.7)$$

which requires that the fulfillment of solvability condition

$$\tilde{f}_2(p,\xi) = -\frac{\beta_0}{\alpha_0}\frac{\tilde{f}_1(p,\xi)}{\lambda(p,\xi)}, \quad \lambda = \sqrt{\frac{p}{a} + \xi^2}, \qquad (2.4.8\alpha)$$

$$\tilde{f}_2(\alpha\xi^2,\xi) = -\frac{\beta_1}{\alpha_1}\frac{\tilde{f}_1(\alpha\xi^2,\xi)}{\sigma(p,\xi)}, \quad \sigma = \sqrt{\frac{b-a}{b-a\chi^2}}, \qquad (2.4.8\beta)$$

$$\tilde{f}_2(p,0) = -\frac{\beta_0}{\alpha_0}\frac{\tilde{f}_1(p,0)}{\lambda_a}, \quad \lambda_a = \sqrt{\frac{p}{a}}. \qquad (2.4.8\gamma)$$

Since $C^n(p,\xi)$ is separated from classical solutions, equalities (2.4.4) are satisfied exactly.

The problem solution for ODE is reconstructed by formula (2.4.3)

$$\tilde{u}_n(p,x,\xi) = C_1^n(p,\xi)\mathrm{e}^{-x\lambda_a}, \quad \tilde{v}_n(p,x,\xi) = C_2^n(p,\xi)\mathrm{e}^{-x\lambda_b}.$$

These equalities determine the ordinary (exact) solution of the problem (α) and nonclassical solutions of the problems (β) and (γ), satisfying the system of differential equations, but at the same time boundary conditions do not always hold because the functions $F(\xi)$ and $F(p)$ are given only along the parabola $p = \alpha\xi^2$, $\alpha > 0$, $\xi \neq 0$ and along the ray $p > 0$, $\xi = 0$, respectively.

Determining of originals

The calculations are based on the tables of formulas presented in Bateman and Erdélyi (1969), Gradstein and Ryzhik (1971).

A. α) Rewriting (2.4.8α) as

$$\tilde{f}_2(p,\xi) = -\sqrt{a}\,\frac{\beta_0 \tilde{f}_1(p,\xi)}{\alpha_0\sqrt{p+a\xi^2}},$$

we obtain

$$f_2(t,y) = -(\beta_0/\alpha_0)\sqrt{a}F_a^{-1/2} f_1(\tau,z). \qquad (2.4.8\alpha')$$

Here

$$F_a^{-1/2} f = \frac{1}{\pi}\int_0^t \int_{-\infty}^{\infty} f(\tau,z)E_a(t-\tau,y-z)\,\frac{\mathrm{d}z\,\mathrm{d}\tau}{\sqrt{t-\tau}}$$

is the fractional integral operator, and $E_a(t,z)$ is the fundamental solution of heat conduction equation $F_a u \equiv u_t - a\partial_z^2 u = 0$. From (2.4.8$\alpha'$), the function $f_1(t,y)$ can be found by means of the fractional differential operator

$$\frac{\beta_0}{\alpha_0} f_1(t,y) = -\frac{1}{\sqrt{a}} F_a^{-1/2} f_2(\tau,z). \qquad (2.4.9\alpha)$$

β) The calculation is performed by nonstandard methods (see Solonnikov's example)

$$\tilde{f}_2(\alpha\xi^2,\xi) = \int_0^\infty f_2(t,y)\mathrm{e}^{-\mu t}\,\mathrm{d}t = f_{2,\mu}(y) \quad\Leftrightarrow\quad F(\xi),$$

where the parameter $\mu = \alpha\xi^2$ is positive. From the right-hand side of equality (2.4.8β) we obtain

$$\left(\frac{\tilde{f}_1}{\xi}\right) \xrightarrow{\Phi_{\bar{\xi}}} \int_0^\infty \bar{f}_1(p,y-z)\varphi(z)\,\mathrm{d}z = \frac{\pi}{2}\int_0^\infty \bar{f}_1(p,y-z)\big|_{p=\alpha\xi^2}\,\mathrm{d}z$$
$$= \frac{\pi}{2}\int_0^\infty \int_0^\infty f_1(t,y-z)\mathrm{e}^{\mu t}\,\mathrm{d}t\,\mathrm{d}z \quad \left(\varphi(z) = \Phi_{\xi}^{-}\left(\frac{1}{\xi}\right) = \frac{\pi}{2}\right).$$

Therefore, the following equality is true:

$$f_2(t,y) = -\frac{\beta_1}{\sigma\alpha_1}\,\frac{\pi}{2}\int_0^\infty f_1(t,y-z)\,\mathrm{d}z, \qquad (2.4.8\beta')$$

which implies

$$\frac{\pi\beta_1}{2\sigma\alpha_1} f_1(t,y) = \partial_y f_2(t,y). \qquad (2.4.9\beta)$$

γ) Similarly, we treat the condition (2.4.8γ)

$$\tilde{f}_2(p,0) = \tilde{f}_2(p,\xi)\big|_{\xi=0} \xrightarrow{\Lambda_p^-} \bar{f}_2(t,\xi)\big|_{\xi=0}$$
$$\triangleq \int_{\mathbb{R}} f_2(t,y)\mathrm{e}^{i\xi y}\big|_{\xi=0}\,\mathrm{d}y = \int_R f_2(t,y)\,\mathrm{d}y = f_2^0(t) \quad \Leftrightarrow \quad F(p).$$

Proceeding from the other side of this condition, we obtain

$$\frac{\tilde{f}_1(p,0)}{\sqrt{p}} \xrightarrow{\Lambda_p^-} \int_0^t \bar{f}_1(\tau,\xi)\,\frac{\mathrm{d}\tau}{\sqrt{\pi(t-\tau)}}\Big|_{\xi=0} = \int_0^t \frac{\mathrm{d}\tau}{\sqrt{\pi(t-\tau)}}\int_{\mathbb{R}} f_1(\tau,y)\,\mathrm{d}y.$$

Their comparison yields

$$f_2(t,y) = -\frac{\beta_0}{\alpha_0}\sqrt{\frac{a}{\pi}}\int_0^t \frac{f_1(\tau,y)}{\sqrt{\pi(t-\tau)}}\,\mathrm{d}\tau = -\frac{\beta_0\sqrt{a}}{\alpha_0}\,J^{1/2}(f_1). \qquad (2.4.8\gamma')$$

Acting by the operator of fractional differentiation $\partial_t^{1/2}$, we find one more relation between the functions

$$\frac{\beta_0}{\alpha_0}\,f_1(t,y) = -\frac{1}{\sqrt{\pi a}}\,\partial_t\int_0^t \frac{f_1(\tau,y)}{\sqrt{(t-\tau)}}\,\mathrm{d}\tau = -\frac{1}{\sqrt{a}}\,\partial_t^{1/2}f_2. \qquad (2.4.9\gamma)$$

The calculation performed enables us to write the following formulas of inversion:

$$\tilde{W}_a(p,x,\xi) = \hat{f}_2(p,\xi)\mathrm{e}^{-\lambda_a x} \quad \Leftrightarrow$$
$$W_a(t,x,y) = -2a\partial_x\int_0^t \mathrm{d}\tau\int_{\mathbb{R}} f_2(\tau,z)E_a(t-\tau,y-z)\,\mathrm{d}z, \qquad (2.4.10\alpha)$$

$$\tilde{H}_a = F(\xi)\mathrm{e}^{-\lambda_a x} \quad \Leftrightarrow$$
$$H_a(t,x,y) = -2a\partial_x\int_0^t \mathrm{d}\tau\int_{\mathbb{R}} f_{2,\mu}(z)E_a(t-\tau,x,y-z)\,\mathrm{d}z, \qquad (2.4.10\beta)$$

$$\tilde{P}_a = \tilde{f}_2(p,0)\mathrm{e}^{-\lambda_a x} \quad \Leftrightarrow$$
$$P_a(t,x,y) = -2a\partial_x\int_0^t f_2^0(\tau)E_a(t-\tau,x)\,\mathrm{d}\tau, \qquad (2.4.10\gamma)$$

where $\lambda_a = \sqrt{p/a+\xi^2}$.

B. Original problems. At the point $\chi = 1$ there holds the problem (α):

$$u_t = a\Delta u, \qquad v_t = a\Delta v \qquad (a = b),$$
$$u(0, x, y) = v(0, x, y) = 0,$$
$$\beta_0 u_x + \beta_1 v_x = (\beta_1/\alpha_1) f_1(t, y),$$
$$\beta_0 u + \beta_1 v = f_2(t, y).$$

Let us write the solution in the form

$$u_n(t, x, y) = \frac{\beta_0}{\Delta}\, w_a(t, x, y), \qquad v_n(t, x, y) = \frac{\beta_1}{\Delta}\, w_a(t, x, y), \qquad (2.4.11\alpha)$$

where $\Delta = \beta_0^2 + \beta_1^2$, the function w_a is determined from $(2.4.10\alpha)$. Obviously, it is sufficient to verify only boundary conditions. Let us calculate directly

$$\beta_0 u_n + \beta_1 v_n = w_a(t, 0, y) = f_2(t, y),$$

$$\begin{aligned}\beta_0 \partial_x u_n + \beta_1 \partial_x v_n &= \partial_x w_a(t, 0, y)\\ &= -2a \int_0^t \mathrm{d}\tau \int_{\mathbb{R}} f_2(\tau, z) \partial_x^2 E_a(t - \tau, x, y - z)\big|_{x=0}\, \mathrm{d}z\\ &= -\frac{1}{\sqrt{a}}\, F_a^{1/2} f_2(\tau, z) = \frac{\beta_0}{\alpha_0}\, f_1(t, y) = \frac{\beta_1}{\alpha_1}\, f_1(t, y).\end{aligned}$$

We use here the formula $(2.4.9\alpha)$.

Thereby, we show that the solution $u_n = u$, $v_n = v$, $(2.4.11\alpha)$ is classical.

$\beta)$ In the interval $1 < \chi < \sqrt{b/a}$, $b > a$ the solution of the problem (2.1.1) can be represented in the form

$$u_n(t, x, y) = \frac{\beta_0}{\Delta}\, H_a(t, x, y), \qquad v_n(t, x, y) = \frac{\beta_1}{\Delta}\, H_b(t, x, y). \qquad (2.4.11\beta)$$

Here, the functions H_a and H_b are defined by $(2.4.10\beta)$.

Let us again verify the boundary conditions. Taking into account the lemma on limiting values, we have

$$\begin{aligned}H_a^0 &= H_a(t, 0, y)\\ &= \lim_{x \to 0} \frac{x}{4a\pi} \int_0^t \mathrm{d}\tau \int_{\mathbb{R}} f_{2,\mu}(z) \exp\Big(-\frac{x^2 + (y - z)^2}{4a(t - \tau)}\Big) \frac{\mathrm{d}z}{(t - \tau)^2} = f_{2,\mu}(y).\end{aligned}$$

Therefore,

$$\beta_0 u_n + \beta_1 v_n = f_{2,\mu}(y) = \int_0^\infty f_2(t, y) \mathrm{e}^{-\mu t}\, \mathrm{d}t, \qquad \mu > 0. \qquad (2.4.12.2)$$

Let us find the limiting value

$$\begin{aligned}
\partial_x H_a(t,x,y)\big|_{x=0} &= \partial_x H_a^0 = -2a\int_0^t \mathrm{d}\tau \int_{\mathbb{R}} f_{2,\mu}(z)\partial_x^2 E_a(t-\tau,0,y-z)\,\mathrm{d}z \\
&= 2a\int_0^t \mathrm{d}\tau \int_{\mathbb{R}} f_{2,\mu}(z)(\partial_x + a\partial_z^2)E_a(\tau,0,y-z)\,\mathrm{d}z \\
&= 2\int_{\mathbb{R}} f_{2,\mu}(z)E_a(\tau,0,y-z)\,\mathrm{d}z + 2a\int_{\mathbb{R}} \partial_z^2 f_{2,\mu}(z)\,\mathrm{d}z \int_0^t E_a(\tau,0,y-z)\,\mathrm{d}\tau \\
&= 2\int_{\mathbb{R}} f_{2,\mu}(z)E_a(\tau,0,y-z)\,\mathrm{d}z,
\end{aligned}$$

since

$$\begin{aligned}
\int_0^t E_a(\tau,0,y)\,\mathrm{d}\tau &= \frac{1}{4\pi a}\int_0^t \mathrm{e}^{-y^2/(4a\tau)}\,\frac{d\tau}{\tau} = \left| \begin{matrix} \dfrac{y^2}{4a\tau} = \eta, \quad \mathrm{d}\tau = -\dfrac{y^2}{4a\eta^2} \\ u(y) = \eta_b = \frac{y^2}{4at} \end{matrix} \right| \\
&= \frac{1}{4\pi a}\int_{u(y)}^t \mathrm{e}^{-\eta}\,\frac{\mathrm{d}\eta}{\eta} = \frac{\mathrm{e}^{-u/2}}{4a\pi}\,\frac{W^{(u)}_{-1/2,0}}{\sqrt{u}} = \frac{\mathrm{e}^{-u/2}}{4a\pi}\,\frac{M^{(u)}_{-1/2,0}}{\sqrt{u}} \\
&= \frac{\mathrm{e}^{-u}}{2a\pi}\,\Phi(1,1;u) = \frac{1}{2a\pi}.
\end{aligned}$$

Further,

$$\begin{aligned}
\partial_x H_a^0 &= 2C\int_{\mathbb{R}} E_a(t-\tau,0,y-z)\,\mathrm{d}z \int_0^\infty f_{1,\mu}(t,z-\eta)\,\mathrm{d}\eta \\
&= 2C\int_{\mathbb{R}} E_a(t-\tau,0,y-z)\,\mathrm{d}z \int_z^\infty f_{1,\mu}(t,\xi)\,\mathrm{d}\xi \\
&= 2C\int_{\mathbb{R}} f_{1,\mu}(t,\xi)\,\mathrm{d}\xi \int_{-\infty}^{\xi} E_a(t,0,y-z)\,\mathrm{d}z \\
&= \frac{C}{2\sqrt{a\pi t}}\int_{\mathbb{R}} f_{1,\mu}(t,\xi)\Big[1-\Phi\Big(\frac{y-\xi}{2\sqrt{at}}\Big)\Big]\,\mathrm{d}\xi = \frac{Cf_2^0(t,y)}{2\sqrt{\pi a t}},
\end{aligned}$$

where

$$\Phi(x) = \operatorname{erf} x = \frac{2}{\sqrt{\pi}}\int_0^x \mathrm{e}^{-t^2}\,\mathrm{d}t$$

is the probability integral, $C = -\pi\beta_1/(\sigma\alpha_1)$.

As a result, the first boundary condition is as follows:

$$\alpha_0\partial_x u_n + \alpha_1\partial_x v_n = \frac{\text{const}}{\sqrt{t}}\, f_1^0(t,y), \tag{2.4.12.1}$$

where

$$\text{const} = -\frac{\sqrt{\pi}}{4\sigma\Delta}\Big(\frac{\beta_0^2}{x\sqrt{a}} + \frac{\beta_1^2}{\sqrt{b}}\Big), \quad f_1^0(t,y) = \int_{\mathbb{R}} \Big[1 - \Phi\Big(\frac{y-\xi}{2\sqrt{at}}\Big)\Big] f_{1,\mu}(t,\xi)\,\mathrm{d}\xi.$$

Therefore, the boundary conditions in this problem turn into inequalities (2.4.12.i), $i = 1, 2$.

γ) At the point $\chi = \sqrt{b/a}$, $b > a$ the boundary conditions take the form

$$\beta_0 u_x + \beta_1 \sqrt{\frac{b}{a}}\, v_x = \frac{\beta_0}{\alpha_0}\, f_1(t,y), \qquad \beta_0 u + \beta_1 v = f_2(t,y).$$

The solution is of the form

$$u_n(t,x,y) = \frac{\beta_0}{\Delta}\, P_a(t,x,y), \qquad v_n(t,x,y) = \frac{\beta_1}{\Delta}\, P_b(t,x,y). \tag{2.4.11γ}$$

On the boundary of the domain $x = 0$, we have

$$\beta_0 u_n + \beta_1 v_n = f_2^0(t) = \int_{\mathbb{R}} f_2(t,y)\,\mathrm{d}y. \tag{2.4.12γ}$$

Let us determine the derivative

$$\begin{aligned}
\partial_x P_a = -2a \int_0^t f_2(\tau)\partial_x^2 E_a(t-\tau,x)\,\mathrm{d}\tau &= 2\int_0^t f_2(\tau)\partial_\tau E_a(t-\tau,x)\,\mathrm{d}\tau \\
= 2\Big[f_2(\tau)E_a(t-\tau,x)\Big|_{\tau=0}^{\tau=t} &- \int_0^t f_2'(\tau)E_a(t-\tau,x)\,\mathrm{d}\tau\Big] \\
&= -2\Big[f_2(0)E_a(t,x) + \int_0^t f_2'(\tau)E_a(t-\tau,x)\,\mathrm{d}\tau\Big]
\end{aligned}$$

and calculate its value for $x = 0$

$$\partial_x P_a\big|_{x=0} = -\Big[\frac{f_2(+0)}{\sqrt{a\pi t}} + \int_0^t \frac{f_2'(\tau)\,\mathrm{d}\tau}{\sqrt{a\pi(t-\tau)}}\Big] = -\frac{1}{\sqrt{a\pi}}\,\partial_t \int_0^t \frac{f_2(\tau)\,\mathrm{d}\tau}{\sqrt{(t-\tau)}}.$$

Now we return to the first boundary equality

$$\begin{aligned}
\beta_0\partial_x u_n + \beta_1\sqrt{\frac{b}{a}}\,\partial_x v_n &= \frac{\beta_0^2}{\Delta}\,\partial_x P_a\big|_{x=0} + \sqrt{\frac{b}{a}}\frac{\beta_1^2}{\Delta}\,\partial_x P_b\big|_{x=0} \\
&= -\frac{1}{\sqrt{a\pi}}\,\partial_t \int_0^t \frac{f_2(\tau)\,\mathrm{d}\tau}{\sqrt{(t-\tau)}} = \frac{\beta_0}{\alpha_0}\, f_1(t,y).
\end{aligned}$$

The last transition is due to formula (2.4.9γ).

Thus, the first boundary condition is satisfied, and the second one assumes the form (2.4.12γ).

Conclusion. Depending on the kind of the function entering into the formulas for the solution $C^n(p,\xi)$ of SLAE, the original problem has an exact solution (the case (α)) or a nonclassical solution, for which not all the equalities of the boundary data are true (the cases (β) and (γ)).

2.4.4. Case of well-posedness ($\det \mathcal{F}(p,\xi) \neq 0$)

Let us use the results of Basarbayeva (1982). For better compactness and simplification of the analysis, the following notation is introduced

$$v_1 = u, \qquad v_2 = v, \qquad \lambda_1 = a, \qquad \lambda_2 = b,$$

$$a_k = \frac{\alpha_{2-k}\beta_{k-1}}{\sqrt{\lambda_k}} \quad (k = 1, 2), \qquad A = \frac{\lambda_1 a_1^2 - \lambda_2 a_2^2}{a_1^2 - a_2^2}.$$

The solution of SLAE (2.4.4) is of the form

$$C_k(p,\xi) = \frac{(-1)^{k-1}(\beta_{2-k}\tilde{f}_1 + \alpha_{2-k}\tilde{f}_2\sqrt{p/\lambda_{3-k} + \xi^2})}{\det \mathcal{F}(p,\xi)}, \qquad k = 1, 2.$$

Let us represent the function in the form more convenient for calculation

$$\tilde{v}_k(p,\xi) = (\beta_{2-k}\tilde{f}_1 + \alpha_{2-k}\tilde{f}_2\sqrt{p/\lambda_{3-k} + \xi^2})\tilde{K}_k(p,x,\xi),$$

where

$$\tilde{K}_k = \frac{p + \lambda_k\xi^2}{\sqrt{p/\lambda_k + \xi^2}}\, \mathrm{e}^{-x\sqrt{p/\lambda_k + \xi^2}}\, \tilde{g}_k(p,\xi),$$

$$\tilde{g}_k(p,\xi) = \frac{(-1)^{k-1}}{\sqrt{\lambda_k}\det \mathcal{F}(p,\xi)\sqrt{p + \lambda_k\xi^2}}, \qquad k = 1, 2.$$

Using the method of contour integration and the Jordan lemma we arrive at

$$\bar{g}_k(t,\xi) = \mu_k \int_{\lambda_1}^{\lambda_2} \frac{\sqrt{|z - \lambda_{3-k}|}}{(z - A)\sqrt{|z - \lambda_k|}}\, \mathrm{e}^{-t\xi^2 z}\, \mathrm{d}z + \gamma_k \mathrm{e}^{-A\xi^2 t},$$

where

$$\mu_k = \frac{a_{3-k}}{\pi(a_1^2 - a_2^2)\sqrt{\lambda_k}}, \qquad \gamma_k = \frac{(-1)^k}{\sqrt{\lambda_k}}\, \frac{a_1|a_2| + a_2|a_1|}{|a_{3-k}|(a_1^2 - a_2^2)}.$$

Continue the calculation

$$g_k(t,\xi) = \mu_k \int_{\lambda_1}^{\lambda_2} \frac{\sqrt{|z - \lambda_{3-k}|}}{(z - A)\sqrt{|z - \lambda_k|}}\, \frac{\mathrm{e}^{-y^2/(4tz)}}{\sqrt{4\pi t z}}\, \mathrm{d}z + \gamma_k \frac{\mathrm{e}^{-y^2/(4At)}}{\sqrt{4\pi A t}},$$

$$K_k(t,x,y) = \frac{\mu_k}{4\pi} \int_{\lambda_1}^{\lambda_2} \varsigma_k(t,x,y;z_1)\, \frac{\sqrt{|z - \lambda_{3-k}|}}{(z_1 - A)\sqrt{|z_1 - \lambda_k|}}\, \mathrm{d}z_1 + \frac{\mu_k}{4\pi}\, \varsigma_k(t,x,y;A),$$

where

$$\varsigma_k(t,x,y;z_1) = \int_0^1 H_k^0(t,x,y;\sigma(z_1,z_2)) \times \exp\Big[-\frac{y^2}{4\sigma_k(z_1,z_2)} - \frac{x^2}{4\lambda_k(1-z_2)t}\Big]\frac{dz_2}{2t(1-z_2)\sqrt{\pi t\sigma_k(z_1,z_2)}},$$

$$H_k^0(t,x,y;\sigma_k) = \frac{\lambda_k}{\sigma_k} + \frac{x^2}{4\lambda_k t} + \Big(1 - \frac{\lambda_k}{\sigma_k}\Big)\frac{y^2}{2t\sigma_k},$$

$$\sigma_k(A,z) = Az + (1-z)\lambda_k, \qquad k = 1,2.$$

Finally, the original is

$$\sqrt{p+\lambda_{3-k}\xi^2}\,\tilde{f}_2(p,\xi) = \frac{p+\lambda_{3-k}\xi^2}{\sqrt{p+\lambda_{3-k}\xi^2}}\,\tilde{f}_2 = F_{\lambda_{3-k}}(F_{\lambda_{3-k}}^{-1/2} f_2(\tau,z)) = F_{\lambda_{3-k}}^{1/2} f_2(\tau,z).$$

Thus, we obtain the sought solution in the form

$$v_k(t,x,y) = \int_0^t d\tau \int_{\mathbb{R}} \Big[\beta_{2-k} f_1(\tau,z) + \frac{\alpha_{2-k}}{\sqrt{\lambda_{3-k}}} F_{\lambda_{3-k}}^{1/2} f_2(\xi,\eta)\Big] \times K_k(t-\tau,x,y-z)\,dz. \quad (2.4.13)$$

It is evident that the calculation is much more complex than in ill-posed problems.

Chapter 3.

Hyperbolic equations

3.1. HYPERBOLICITY AND MIXED PROBLEM

3.1.1. General case

Let us consider linear differential operator in the domain $D \subseteq \mathbb{R}^n$

$$A = \sum_{|\alpha| \leq m} a_\alpha(x)\partial^\alpha \tag{3.1.1}$$

and the corresponding differential equation

$$Au = f. \tag{3.1.2}$$

The principal symbol of the operator A is defined by the formula

$$a_m(x, \xi) = \sum_{|\alpha| = m} a_\alpha(x)\xi^\alpha.$$

Remark 3.1.1. Hyperbolicity is usually determined in the presence of a distinguished variable t or at least the distinguished direction ν.

Definition 3.1.1. The operator A (3.1.1) and equation (3.1.2) are called hyperbolic in the direction ν (at the point x) if,

$$a_m(x, \nu) \neq 0, \tag{3.1.1ν}$$

and for every vector $\xi \in \mathbb{R}^n$ (not proportional to v) all m λ-roots of the equality

$$a_m(x, \xi + \lambda\nu) = 0 \qquad (3.1.2\nu)$$

are real.

Remark 3.1.2. The inequality $(3.1.1\nu)$ implies that the direction ν is not characteristic and vice versa, the equality $(3.1.2\nu)$ corresponds to the characteristicity condition.

One class of hyperbolic equations is of particular interest:

Definition 3.1.2. The operator A and equation (3.1.2) are called strictly hyperbolic in the direction ν (at the point x), if all the λ-roots of the equation $(3.1.2\nu)$ are not only real but also different.

If a system with the matrix operator A $(N \times N)$ is considered, then for it to be hyperbolic the following relations should hold:

$$\det a_m(x, \nu) \neq 0, \qquad \det a_m(x,\ \xi + \lambda\nu) = 0. \qquad (c\nu)$$

3.1.2. *t*-hyperbolicity

We are interested in the operators with the distinguished variable t

$$A = \sum_{|\alpha|+j\leq m} a_{\alpha,j}(t, x)\partial_x^\alpha \partial_t^j, \qquad t \in \mathbb{R}, \quad x \in \mathbb{R}^n. \qquad (3.1.1')$$

Definition 3.1.3. The operator A $(3.1.1')$ is called hyperbolic at the point (t, x) if

$$a_{0,m}(t, x) \neq 0 \text{ (the axis } Ot \text{ is non-characteristic)} \qquad (3.1.1t)$$

and for every $\xi \in \mathbb{R}^n$ all λ-roots of the equation

$$A_m(t, x; \lambda, \xi) \equiv \sum_{|\alpha|+j=m} a_{\alpha,j}(t, x)\xi^\alpha \lambda^j = 0 \qquad (3.1.2t)$$

are real.

Let us consider a 1st order system

$$Lu \equiv AU_t + \sum_{j=1}^{n} B_j U_{x_j} + CU(t, x) = f(t, x) \qquad (3.1.3)$$

with constant square matrices of the size N.

Definition 3.1.4. A system of N first-order equations is called t-hyperbolic if its characteristic equation

$$\det \|\lambda A + \xi B\| = 0 \tag{Ct}$$

for every $\xi \in \mathbb{R}^n$ has N real different λ-roots.

In particular, if all the matrices A and B_j are symmetric and in addition, $A > 0$, then system (2.1.3) is symmetrically hyperbolic (hyperbolic according to Friedrichs).

Recall that any two symmetric matrices A and B (it is admissible that $B = \sum_{j=1}^{n} \xi_j B_j$), one of which (say, A) is positively defined, can be brought to a diagonal form by one and the same nondegenerate real transform (with the matrix V, $\det V \neq 0$) (A can be turned into the unitary matrix $I^{(N)}$).

We consider the system

$$U_t + \sum_{j=1}^{n} B_j U_{x_j} + CU = f(t,x). \tag{3.1.3$'$}$$

The condition of its hyperbolicity (strict hyperbolicity) implies that for every $\xi \in \mathbb{R}^n$ all eigenvalues of the matrix

$$B = \sum_{j=1}^{n} \xi_j B_j \tag{3.1.2$'t$}$$

are real (and different).

3.1.3. Problem statement

Let t be a hyperbolic system given in the canonical form

$$AU_t + \Lambda U_X + \sum_{j=1}^{n} B_j U_{y_j} + QU = 0, \tag{3.1.3$''$}$$

where all the matrices are symmetric, in addition, $A > 0$ (possibly is unitary)

$$\Lambda = \operatorname{diag}(\lambda_1, \dots, \lambda_k; -\lambda_{k+1}, \dots, -\lambda_{k+m}; 0, \dots, 0),$$
$$1 \leq k \leq N, \qquad k + m + r = N, \qquad 0 \leq m, \quad r < N$$

(there are k positive numbers, m negative ones, and r zero numbers).

Problem. It is required to find the bounded solution of system (3.1.3″) in the quadrant $D_{n+2} = (t > 0,\ x > 0,\ y \in \mathbb{R}^n,\ n \geq 0)$ with the initial condition $(t = 0)$

$$U(0, x, y) = f(x, y), \qquad x > 0, \quad y \in \mathbb{R}^n \tag{3.1.4}$$

and the boundary condition $(x = 0)$

$$MU(t, 0, y) = \varphi(x, y), \qquad t > 0, \quad y \in \mathbb{R}^n. \tag{3.1.5}$$

It is required that on every boundary of the domain (with a separating boundary) the number of imposed conditions corresponds to the number of outgoing characteristics of the system of equations (Godunov, 1971). For the left boundary $(x = 0)$ the characteristics with the positive slope $\lambda_j > 0$ are " outgoing". Consequently, the matrix M should have k rows; the right boundary of this domain goes into infinity, therefore the bounded solution is sought; from the boundary $t = 0$ there "outgo" vertical $(\lambda_j = 0)$ characteristics and in order to find the components $U_{k+m+1}, \ldots, U_N$ it is sufficient to confine ourselves to the initial data.

Thus, boundary condition (2.1.5) in this standard statement would be better rewritten as

$$U^{\mathrm{I}} = LU^{\mathrm{II}} + U(t, y), \tag{3.1.5′}$$

where the vectors $U^{\mathrm{I}} = (U_1, \ldots, U_k)$, $U^{\mathrm{II}} = (U_{k+1}, \ldots, U_{k+m})$, and the matrix $L = (a_{ij})$, $i = 1, \ldots, k$, $j = k+1, \ldots, k+m$, is constant, otherwise $M = (I^{(k)}, -L)$ is of the order $k \times (N - r)$.

Most probably in the above statement problem (2.1.3″), (2.1.4), (2.1.5′) will always be well-posed. Therefore, we will be interested in other forms of specifying the boundary conditions. That is, for the matrix M we retain the dimensionality but change its structure, assuming that M is constant or $M(\lambda)$ is differential (of an arbitrary order) with constant coefficients.

According to the structure of the matrix Λ, we arrive at the equalities

$$\dot{v} = Av + F, \qquad A_1 v = F_1, \qquad x > 0,$$
$$Mv = \varphi, \qquad x = 0.$$

From the algebraic part of equations we find the components $v_{k+m+1}, \ldots, v_N$, and then substitute them into the differential part and into the boundary conditions thus obtaining the boundary-value problem

$$\dot{W} = AW + \psi, \qquad x > 0,$$
$$BW(0) = \psi,$$

where the vector $W = W(x) = W(p,x,\omega) = (W_1, \dots, W_{k+m})$, the matrix $A = A(p,\omega)$ is a square matrix of the order $(k+m)$, $B = B(p,\omega)$ is rectangular $k \times (k+m)$ matrix. The vector W and matrices A, B depend on parameters of the Fourier (Laplace) integral transforms.

Further, we study well-known physical models.

3.2. TWO-DIMENSIONAL ACOUSTIC PROBLEM

Consider the system describing the propagation of sound waves in the medium at rest for the two-dimensional case $t > 0$, $x > 0$, $y \in \mathbb{R}$ (Godunov, 1971)

$$\begin{aligned} (\rho M^2)^{-1} d_t + u_x + v_y &= 0, \\ \rho u_t + d_x &= 0, \\ \rho v_t + d_y &= 0, \end{aligned} \tag{3.2.1}$$

where u, v are components of the vector $U = (d,u,v)$ denoting the velocities of perturbed medium; d is the pressure. The constants ρ, $M > 0$ are connected with the properties of the medium at rest: ρ is the density, M is compressibility.

Problem. Find the solution of system (3.2.1) given the initial condition on the plane $t = 0$, $x \geq 0$, $y \in \mathbb{R}$

$$u(0,x,y) = f(x,y) \tag{3.2.2}$$

and the boundary condition at $x = 0$, $t \geq 0$, $y \in \mathbb{R}$

$$lu \equiv d + a\rho u + b\rho v = \varphi(t,y), \qquad ab \neq 0. \tag{3.2.3}$$

For existence of regular solution, it is required that the consistency condition is fulfilled between the vector f and the function φ:

$$f_1(0,y) + a\rho f_2(0,y) + b\rho f_3(0,y) = \varphi(0,y). \tag{3.2.4}$$

Many authors had earlier studied this problem and several interesting facts were discovered. Some of them are given below.

The domain of ill-posedness O is determined by Godunov and Gordiyenko (1977), as the authors themselves put it, by applying the results obtained by Sakamoto (1972) on energy inequalities for general hyperbolic mixed problems to the boundary-value problem for the wave equation.

In doing this, they discovered the possibility to formulate the ill-posedness condition in terms of the original problem.

Although the Hadamard example for the domain O was constructed by Godunov and Gordiyenko (1977), Fedosov and Yanenko (1982) investigated the character (type) of ill-posedness. It was discovered that in a part of the domain O there is ill-posedness of Cauchy-problem type, and in other part, there is ill-posedness of the Cauchy problem for an elliptical equation, and the lines of parabolic degeneration are given.

The ill-posed case of the problem is studied by Temirbulatov (1989a) using modified potential method. In Temirbulatov (1993) the same problem is investigated at ODE level.

Finally, note that outside the domain O, energy inequalities were obtained by Godunov and Gordiyenko (1977), and the solution itself was found by Fedosov and Yanenko (1982).

This brief digression shows that even the simplest acoustic problems are very interesting and their investigation can bring about important results.

3.2.1. Reduction of (3.2.1)–(3.2.3) to new problems

Integral Fourier $F_y^+(\omega)$ and Laplace $\Lambda_t^+(p)$ transforms yield a boundary-value problem with respect to the images:

$$(\hat{d}, \hat{u})\,(p, x, \omega) \equiv w(p, x, \omega) = (w_1, w_2)(p, x, \omega) = w(x)$$

for the first-order ODE

$$\dot{w}(x) = A(p, \omega) w(x) + \tilde{F}(x, \omega), \qquad x > 0, \tag{3.2.5}$$

$$(1 - i\omega b/p) w_1 + a\rho w_2 = \tilde{\varphi}(p, \omega) - b\rho \tilde{f}_3(0, \omega)/p, \qquad x = 0. \tag{3.2.6}$$

The third component is expressed explicitly

$$\hat{v}(x) = w_3(x) = (\tilde{f}_3(x, \omega) - i\omega w_1(x)/\rho)/p. \tag{3.2.7}$$

Here and below the following denotation is used

$$A(\rho, \omega) = \begin{bmatrix} 0 & -\rho p \\ -r^2/(\rho M^2 p) & 0 \end{bmatrix}, \qquad \tilde{F} = \left\| \begin{matrix} \rho \tilde{f}_2 \\ \tilde{f}_1/(\rho M^2) - i\omega \tilde{f}_3/p \end{matrix} \right\|,$$

$$r = \sqrt{p^2 + M^2 \omega^2}, \qquad \operatorname{Re} r > 0,$$

$$\tilde{f}_j = \tilde{f}_j(x, \omega) = F_y^+ f_j(x, y) = \int_{\mathbb{R}} f_j(x, y) e^{i\omega y}\, dy.$$

For the matrix $A(p,\omega)$, we find the eigenvalues $\lambda = \pm r/M$ and eigenvectors $z^{\pm} = (1, \mp r/(\rho M p))$ corresponding to them.

By means of $V = (z^{-}, z^{+})$ we construct the Green matrix of system (3.2.5) (Godunov, 1994)

$$G_0(x,A) = \begin{cases} V\begin{pmatrix} e^{\lambda^- x} & 0 \\ 0 & 0 \end{pmatrix} V^{-1} = \dfrac{e^{\lambda^- x}}{2}\begin{bmatrix} 1 & R^{-1} \\ R & 1 \end{bmatrix} & x > 0, \\ V\begin{pmatrix} 0 & 0 \\ 0 & e^{\lambda^- x} \end{pmatrix} V^{-1} = \dfrac{e^{\lambda^+ x}}{2}\begin{bmatrix} -1 & R^{-1} \\ R & -1 \end{bmatrix} & x < 0, \end{cases}$$

where $R = r/(\rho M p)$.

Determine the particular solution of inhomogeneous system (3.2.5)

$$w^1(x) = \frac{1}{2}\int_0^x e^{\lambda^-(x-s)} \begin{bmatrix} \chi^+ \\ \theta^+ \end{bmatrix} (p,s,\omega)\, ds + \frac{1}{2}\int_x^\infty e^{\lambda^-(x-s)} \begin{bmatrix} \chi^- \\ \theta^- \end{bmatrix} (p,s,\omega)\, ds, \quad (3.2.8)$$

where

$$\chi^{\pm} = \frac{p}{Mr}\tilde{f}_1 - \frac{i\omega\rho M}{r}\tilde{f}_3 \pm \rho \tilde{f}_2, \qquad \theta^{\pm} = \frac{r}{Mp}\tilde{f}_2 \pm \Big(\frac{\tilde{f}_1}{\rho^2 M} - \frac{i\omega}{p}\tilde{f}_3\Big).$$

The solution of the problem bounded on the ray $x > 0$ is sought in the form of the sum

$$w(x) = C(p,\omega)e^{\lambda^- x} z^- + w^1(x), \quad (3.2.9)$$

where $C(p,\omega)$ is an unknown coefficient.

The boundary condition (3.2.3) implies

$$\mathcal{F}(p,\omega)C(p,\omega) = g(p,\omega) + g_0(p,\omega), \quad (3.2.10)$$

where

$$\mathcal{F}(p,\omega) = 1 - \frac{ib\omega}{p} + \frac{ar}{Mp}, \qquad g(p,\omega) = \hat{\varphi}(p,\omega) - b\frac{\rho}{p}\tilde{f}_3(a,\omega),$$
$$g_0(p,\omega) = \frac{1}{2}\int_0^\infty e^{-rs/M}\chi^-(p,s,\omega)\Big(1 - \frac{ib\omega}{p} - \frac{ar}{Mp}\Big)\, ds. \quad (3.2.11)$$

3.2.2. Ill-posedness

Zeroes of the function $\mathcal{F}(p,\omega)$ are studied depending on the numbers a, b and parameters p, ω. We have $p - i\omega b = -ar/M$, whence there follows the necessary condition

$$a < 0. \tag{3.2.12}$$

By squaring the both sides of this equality we arrive at the equation

$$(1 - a^2/M^2)p^2 - 2ib\omega p - (a^2 + b^2)\omega^2 = 0,$$

the roots of which are

$$p_1 = (\gamma + i\beta)\omega, \qquad p_2 = (-\gamma + i\beta)\omega,$$

where $\gamma = -a\sigma\sqrt{D} > 0$, $\beta = b\sigma M$, $\sigma = M/(M^2 - a^2)$.

We can now draw the necessary conclusions

1) Taking into account (3.2.12) we find the set of zeroes

$$P_0 = \begin{cases} \omega = 0,\ p \text{ is arbitrary}, & a = -M, \\ p = p_1 = \alpha\omega,\ \omega > 0,\ \alpha = \gamma + i\beta, & a^2 + b^2 < M^2,\ a < 0. \end{cases}$$

Here the root $p = p_1$ is chosen because for it ($\operatorname{Re} p_1 > 0$) the Hadamard example can easily be constructed (Ladyzhenskaya, Uraltseva, and Solonnikov, 1967), and if $\operatorname{Re} p \leq 0$, then, as is known, the solution u is expressed in terms of $\hat{u}$ using integral transform inverse to Φ_y^+ and Λ_t^+ (see also Temirbulatov, 1989a).

2) The domains of ill-posedness are the straight line $a + M = 0$ and the semicircle $a^2 + b^2 < M^2$, $a < 0$ on the plane (a, b).

3) There are two types of the H-condition of ill-posedness:

β) "root": the parameters are connected by the relations $p = \alpha\omega$, $\omega > 0$;

γ) "coefficient-root": $a = -M$ in this case $\omega = 0$ (ω is fixed), and the other parameter p is arbitrary.

3.2.3. The inverse approach. Problem solution with the I-condition

Let us denote the problem by symbols β or γ.

Step 1. Algebraic equation (the condition of solvability, calculation of the coefficient C^l). Let us formally find from (3.2.10) the quantity

$$C(p,\omega) = \hat{C}(p,\omega)/N(p,\omega), \tag{3.2.13}$$

where $\hat{C}(p,\omega) = p\hat{\varphi}(p,\omega) - bp\tilde{f}_3(0,\omega) + pg_0(p,\omega)$, $N(p,\omega) = p - i\omega b + ar/M$.

Since on the set P_0 the denominator of fraction (3.2.13) tends to zero, we choose the function $\hat{\varphi}(p,\omega)$ and vector $\tilde{f}(x,\omega)$ such that the numerator on P_0 is also equal to zero. By using L'Hopital's rule, we evaluate the obtained indeterminate form.

γ) Thus, when $a=-M$, assuming $\hat{C}(p,\omega)=0$, we obtain

$$C^l < (p,\omega) = C_\gamma(p) = \lim_{\varpi\to 0}\frac{\partial_\omega \hat{C}(p,\omega)}{-ib+\omega M^2/r} = \frac{ip}{b}\,\partial_\omega\hat{\varphi}(p,0) - ip\partial_\omega \tilde{f}_3(0,0)$$
$$+\frac{ip}{bM}\int_0^\infty \partial_\omega(\tilde{f}_1 - pM\tilde{f}_2)(s,0)\mathrm{e}^{-ps/M}\,\mathrm{d}s + \frac{1}{2}\int_0^\infty \tilde{\Phi}(s,0)\mathrm{e}^{-ps/M}\,\mathrm{d}s, \tag{3.2.14}$$

where $\tilde{\Phi}(s,0)=\tilde{f}_1/M - p\tilde{f}_2 + 2pM\tilde{f}_3/b$.

Let us write the condition of solvability $\hat{C}(p,0)=0$:

$$p\hat{\varphi}(p,0) - b\rho\tilde{f}_3(0,0) = \frac{p}{M}\int_0^\infty \mathrm{e}^{-ps/M}(\rho M\tilde{f}_2 - \tilde{f}_1)(s,0)\,\mathrm{d}s. \tag{3.2.15γ}$$

β) If $p=\alpha\omega$, then $\hat{C}(\alpha\omega,\omega)=0$, therefore

$$\hat{\varphi}(\alpha\omega,\omega) - \frac{bp}{\alpha\omega}\tilde{f}_3(0,0) + g_0(\alpha\omega,\omega) = 0. \tag{3.2.15β}$$

Let us consider the equation

$$g_0(\alpha\omega,\omega) = \frac{\rho}{2n\alpha}\int_0^\infty \mathrm{e}^{-\mu s}\tilde{F}(s,\omega)\,\mathrm{d}s, \qquad \mu=\frac{|\omega|}{M},$$

and decompose the function $\tilde{F}$ on the real and imaginary parts:

$$\operatorname{Re}\tilde{F} = \sum_{j=1}^{3} c_j\tilde{f}_j, \qquad \operatorname{Im}\tilde{F} = \sum_{j=1}^{3} d_j\tilde{f}_j,$$

where

$$c_1 = [\gamma^2-\beta^2+b\beta - a(\gamma n_0 - \beta n_1)/M]/(M\rho),$$
$$c_2 = a(\gamma^2-\beta^2)/M + (\beta-b)n_1 + aM - \gamma n_0,$$
$$c_3 = an_0 - \gamma M, \qquad d_1 = [(2\beta-b)\gamma - a(\beta n_0+\gamma n_1)/M]/((M\rho),$$
$$d_2 = 2a\gamma\beta/M - (\beta-b)n_0 - \gamma n_1, \qquad d_3 = an_1 - \beta M,$$
$$n=\sqrt{a^2+M^2} = n_0 + in_1, \qquad n_0 = M\sigma\sqrt{D}, \qquad n_1 = -ab\sigma.$$

Let us calculate the sought coefficient

$$C^l < (p,\omega) = C_\beta(\omega) = \lim_{x\to\alpha} \frac{x\hat\varphi(x\omega,\omega) - b\rho\tilde f_3(0,\omega)/\omega + xg_0(x\omega,\omega)}{x - ib + a\sqrt{x^2+M^2}/M}$$
$$= \frac{Mn}{Mn + a\alpha}\Big[\frac{b\rho}{a\omega}\,\tilde f_3(0,\omega) + \alpha\partial_p(\hat\varphi + g_0)(\alpha\omega,\omega)\Big], \quad (3.2.16)$$

where

$$\partial_p g_0(\alpha\omega,\omega) = \frac{1}{2n}\int_0^\infty \mathrm{e}^{-\mu s}\Big\{-\frac{\rho s}{Mn}\,\tilde F(s,\omega) + \frac{1}{\omega}\Big[\frac{M}{n^2}\,\tilde f_1 - \frac{a\rho M}{\alpha^2}\,\tilde f_2$$
$$+ \frac{b\rho M}{\alpha^2 n^2}(n^2+a^2)\tilde f_3 - i\,\frac{\rho}{n^2}\,\mathrm{Im}\,\tilde F - \frac{irM^2}{n\alpha^2}(b\tilde f_2 + a\tilde f_3)\Big]\Big\}(s,\omega)\,\mathrm{d}s.$$

This proves

Lemma 3.2.1. *On the straight line $a+M=0$ provided that $\hat C(p,0)=0$ coefficient $C^l(p,\omega)$ is equal to $C_\alpha(p)$ and determined by (3.2.14).*

In the semicircle $a^2+b^2<M^2$, $a<0$ assuming $\hat C(\alpha\omega,\omega)=0$, we find by formula (3.2.16) $C^l < (p,\omega) = C_\beta(\omega)$.

It is clear that the values $C_\alpha(p)$ and $C_\beta(\omega)$ are inexact solutions of algebraic equation (3.2.10).

Step 2. Solution of the problem for ODE. Let us reconstruct the vector $w(x)$ using formulas (3.2.9) and (3.2.7)

$$W(x) = C(p,\omega)\mathrm{e}^{-rx/M}\begin{bmatrix} 1 \\ r/(\rho Mp) \\ -i\omega/(\rho p)\end{bmatrix} + \begin{bmatrix} w_1^1(x) \\ w_2^1(x) \\ \tilde f_3(x,\omega)/p - i\omega w_1^1(x)/(\rho p)\end{bmatrix}$$
$$= C(p,\omega)\hat q_j(p,x,\omega) + w_j^1(p,x,\omega), \quad (3.2.17)$$

where the coefficient $C(p,\omega)$ is determined by Lemma 3.2.1, the components w_j^1, $j=1,2$ by formula (3.2.8).

Note that "solution" (3.2.17) satisfies system (3.2.5), and boundary conditions (3.2.6) are satisfied only where the functions $C_\alpha(p)$ and $C_\beta(\omega)$ are defined, i. e., on the half-plane $\mathrm{Re}\,p>0$, $\omega=0$ and above the half-plane $p=\alpha\omega$, $\omega>0$, respectively.

Step 3. Solution of the original problem. Recall again one of the main aspects of the method employed. The adverse effect of the set P_0 (or the I-condition of ill-posedness) is taken into account in Step 1 on calculating the "constants" $C_\alpha(p)$ and $C_\beta(\omega)$. Also in Step 1 and the solvability conditions are formulated in the form of restriction on the class of functions $\{\tilde{f}(x,\omega), \hat{\varphi}(p,\omega)\}$ and as a result the parameters p and ω become arbitrary. Thereby the obstacle to the inverse transform of the functions is removed.

Determining originals. Let us employ the rules of operational calculus (especially convolution theorems) and the tables of integral transformations (Bateman and Erdélyi, 1969; Gradstein and Ryzhik, 1971).

Taking into account (3.2.17), by virtue of well-known theorems on convolutions of two functions, we obtain

$$\begin{aligned} u_j(t,x,y) &= (C * q_j) + u_j^1(t,x,y) \\ &= \int\!\!\int C(\tau,z) q_j(t-\tau, x, y-z)\,\mathrm{d}z\,\mathrm{d}\tau + u_j^1(t,x,y). \end{aligned} \tag{3.2.18}$$

a) The components

$$\hat{q}_1 = \mathrm{e}^{-rx/M} = -M\partial_x \hat{q}_0(p,x,\omega), \qquad \hat{q}_0 = \mathrm{e}^{-rx/M}/r,$$

$$\hat{q}_2 = \frac{1}{\rho M}\Big(\frac{r}{p}\,\mathrm{e}^{-rx/M}\Big) = \frac{M}{\rho}\,\partial_x^2 \frac{\hat{q}_0}{p},$$

$$\hat{q}_3 = \frac{1}{\rho}\Big(\frac{i\omega r}{p}\,\mathrm{e}^{-rx/M}\Big) = -\frac{i\omega M}{\rho}\,\partial_x \frac{\hat{q}_0}{p}.$$

As we can see, it is sufficient to transform the image $\hat{q}_0$, then all the rest can be obtained from it by applying integration and differentiation rules for originals.

Thus,

$$\hat{q}_0 = \frac{\mathrm{e}^{-rx/M}}{r} \overset{\Lambda_p^-}{\Rightarrow} \tilde{q}_0(t,x,\omega) = \begin{cases} J_0(\nu\omega), & x < Mt,\ \nu = \sqrt{M^2t^2 - x^2}, \\ 0, & Mt \le x. \end{cases}$$

According to the similarity property

$$J_0(\nu\omega) \overset{\Phi_\omega^-}{\Rightarrow} \quad q_0(t,x,y) = \frac{1}{\pi\sqrt{M^2t^2 - x^2 - y^2}}, \qquad |y| < \nu.$$

Finally, we have

$$\begin{aligned} q_1(t,x,y) &= -M\,\partial_x q_0(t,x,y), \\ q_2(t,x,y) &= \frac{M}{\rho}\,\partial_x^2 \int_0^t q_0(\tau,x,y)\,\mathrm{d}\tau, \\ q_3(t,x,y) &= \frac{M}{\rho}\,\partial_{xy}^2 \int_0^t q_0(\tau,x,y)\,d\tau. \end{aligned} \tag{3.2.19}$$

The support of the function $q_0(t,x,y)$ is a part of the surface of a secular cone with the apex at the point $(t,0,0) \Rightarrow K : \{x^2+y^2 < M^2t^2,\ x>0\}$. Thus, we have

$$u_j(t,x,y) = \iint_K C(\tau,z)q_j(t-\tau,x,y-z)\,dz\,d\tau + u_j^1(t,x,y), \qquad (3.2.18')$$

where

$$d^0 = -M\,\partial_x w_0,$$

$$u^0 = \frac{M}{\rho}\,\partial_x^2 \int_0^t w_0(\xi,x,y)\,d\xi, \qquad v^0 = \frac{M}{\rho}\,\partial_{xy}^2 \int_0^t w_0(\xi,x,y)\,d\xi,$$

$$w_0(t,x,y) = \iint_K C(\tau,z)q_0(t-\tau,x,y-z)\,dz\,d\tau. \qquad (3.2.20\gamma.0)$$

b) Original of the component $\hat{w}^1(x)$. Formulas (3.2.7) and (3.2.8) imply

$$\hat{v}^1 = \frac{1}{p}\Big(\hat{f}_3(x,\omega) - \frac{i\omega}{\rho}\,\hat{d}^1\Big),$$

$$\hat{d}^1 = \frac{1}{2}\int_0^x \Big[\frac{1}{M}\tilde{f}_1\cdot p\hat{q}_0 - \rho M\tilde{f}_3\cdot i\omega\hat{q}_0 + \rho\tilde{f}_2\cdot\hat{q}_1\Big](p,x-s,\omega)\,ds$$
$$+\frac{1}{2}\int_0^x \Big[\frac{1}{M}\tilde{f}_1\cdot p\hat{q}_0 - \rho M\tilde{f}_3\cdot i\omega\hat{q}_0 - \rho\tilde{f}_2\cdot\hat{q}_1\Big](p,s-x,\omega)\,ds,$$

$$\hat{u}^1 = \frac{1}{2}\int_0^x \Big[\frac{1}{M}\tilde{f}_2\cdot\hat{q}_2 + \frac{1}{\rho M^2}\tilde{f}_1\cdot\hat{q}_1 - \tilde{f}_3\cdot\hat{q}_3\Big](p,x-s,\omega)\,ds$$
$$+\frac{1}{2}\int_0^x \Big[\frac{1}{M}\tilde{f}_2\cdot\hat{q}_2 - \frac{1}{\rho M^2}\tilde{f}_1\cdot\hat{q}_1 + \tilde{f}_3\cdot\hat{q}_3\Big](p,s-x,\omega)\,ds.$$

Hence, using the procedure of calculating $q_j(t,x,y)$ (see (3.2.19)), we obtain

$$d^1(t,x,y) = \partial_t M(f_1) - \rho M^2(\partial_x M(f_2) + \partial_y M(f_3)),$$

$$u^1(t,x,y) = M^2\int_0^t \Big(\partial_x^2 M(f_2) + \partial_{xy}^2 M(f_3) - \frac{1}{\rho}\,\partial_x M(f_3)\Big)\,d\tau,$$

$$v^1(t,x,y) = f_3(x,y)\theta(t) - \frac{1}{\rho}\,\partial_y\int_0^t d^1(\tau,x,y)\,d\tau, \qquad (3.2.20\gamma.1)$$

where

$$M(f) = \frac{1}{2M}\iint_d f(s,z)q_0(\tau,x-s,y-z)\,ds\,dz$$
$$= \frac{1}{2\pi M}\iint_d \frac{f(s,z)}{\sqrt{M^2t^2-(s-x)^2-(y-z)^2}}\,ds\,dz.$$

Integration is performed over the interior of the disk $d \equiv (s-x)^2+(y-z)^2 \le M^2t^2$ of the radius Mt with the center at the point (x,y).

From formulas (3.2.20γ.0) and (3.2.20γ.1) there immediately follow the relations

$$u_t = -d_x/\rho, \qquad v_t = -d_y/\rho,$$

i. e., for the vector $U = (d, u, v)$ two last equations of system (3.2.1) are fulfilled. On their basis, the first equation can be readily reduced to the wave equation

$$\Box d \equiv (\partial_t^2 - M^2\Delta)d = 0.$$

Temirbulatov (1980) shown that the function $w_0(t, x, y)$ is a solution to the problem

$$\Box w_0 = 0,$$
$$w_0(0, x, y) = \partial_t w_0(0, x, y) = 0, \qquad \partial_x w_0(t, 0, y) = -C(t, y).$$

Therefore, for the summand $U^0 = (d^0, u^0, v^0)$ equations (3.2.1) are true as well as the zero initial data for every function $C(t, y)$.

In the course of equations of mathematical physics by Bitsadze (1976) the integral expression

$$u(t, x, y) = M(\varphi) + \partial_t M(\psi) \tag{3.2.21}$$

is called Poisson's formula representing the solution to the Cauchy problem

$$\Box u = 0,$$
$$u(0, x, y) = \varphi(x, y), \qquad u_t(0, x, y) = \psi(x, y). \tag{3.2.22}$$

From the last equality of (3.2.20)γ.1) it immediately follows that

$$v^1\big|_{t=0} = f_3(x, y).$$

Further, taking into account (3.2.21), (3.2.22) we obtain

$$d^1\big|_{t=0} = f_1(x, y),$$
$$u^1\big|_{t=0} = M^2 \int_0^t \partial_x^2 M(f_2)\, \mathrm{d}\tau\Big|_{t=0} = f_2(x, y).$$

In the latter one we used the chain of formulas

$$M^2 \int_0^t \partial_x^2 M(f_2)\, \mathrm{d}\tau = \int_0^t (\partial_\tau^2 - M^2\partial_y^2) M(f_2)\, \mathrm{d}\tau$$
$$= \partial_t M(f_2) - M^2\partial_y^2 \int_0^t M(f_2)\, \mathrm{d}\tau.$$

Therefore (d^1, u^1, v^1) satisfies system (3.2.1) and inhomogeneous initial condition (3.2.2).

Let us combine the following two statements:

Lemma 3.2.2. *The functions* $u_j(t,x,y)$ (3.2.18′) *exactly satisfy system* (3.2.1) *and initial condition* (3.2.2).

It remains only to verify boundary condition (3.2.3).

c) The multiplier $C(t,y)$. Let us clarify the following. First, the original of the coefficient $\hat{C}(p,\omega)$ depends on the additional H-condition, represented as (β) and (γ). Second, in order to evaluate the boundary effect assume in initial condition $f\equiv 0$ (further reduced problem (3.2.1)–(3.2.3) is called problem with inhomogeneous boundary condition). This will also help us to avoid unnecessary technical calculations.

Thus, for problem with inhomogeneous boundary condition in the case (γ) solution (3.2.17) is of the form

$$w(x)=\frac{1}{b}\,[ip\partial_\omega\tilde{\varphi}(p,0)\cdot\hat{q}_j(p,x,\omega)], \tag{3.2.17$'\gamma$}$$

and the solvability condition $\tilde{\varphi}(p,0)=0$ is true.

1. Let us make preliminary calculations:

$$ip\,\partial_\omega\hat{\varphi}(p,0)\overset{\Lambda_p^-}{\Rightarrow} i[\partial_\omega\tilde{\varphi}(0,0)+\partial_t\hat{\varphi}_\omega(t,0)],$$

where

$$i\partial_\omega\tilde{\varphi}(0,0)=i\partial_\omega\tilde{\varphi}(0,\omega)\big|_{\omega=0}\overset{\bar{\Delta}}{=}\int_{\mathbb{R}}y\varphi(0,y)e^{i\omega y}\big|_{\omega=0}\,dy=\int_{\mathbb{R}}y\varphi(0,y)\,dy=C_0.$$

This implies that

$$\Lambda_p^-\Phi_\omega^-[ip\partial_\omega\hat{\varphi}(p,0)]=C_0+\partial_t\int_{\mathbb{R}}y\varphi(t,y)\,dy=C_0+\psi'(t),$$

where $\psi(0)=C_0$. As a particular case of the latter we have

$$\Lambda_p^-\Phi_\omega^-[\tilde{\varphi}(p,0)]=\int_{\mathbb{R}}\varphi(t,y)\,dy.$$

On the basis of these calculations and according to formulas (17), let us write out $u_j(t,x,y)$:

$$d=-(M/b)\partial_x(w_0+w)(t,x,y),$$

$$u=\frac{M}{b\rho}\,\partial_x^2\int_0^t(w_0+w)(\xi,x,y)\,d\xi,\quad v=\frac{M}{b\rho}\,\partial_{xy}^2\int_0^t(w_0+w)(\xi,x,y)\,d\xi, \tag{3.2.20$'\gamma$}$$

where

$$w_0 = C_0 \iint_K q_0(t-\tau, x, y-z)\,\mathrm{d}z\,\mathrm{d}\tau,$$
$$w = \iint_K \psi'(\tau) q_0(t-\tau, x, y-z)\,\mathrm{d}z\,\mathrm{d}\tau.$$

These formulas are true only when the solvability condition is satisfied

$$\int_{\mathbb{R}} \varphi(t,y)\,\mathrm{d}y = 0.$$

Let us repeat the integration

$$\begin{aligned} w_0 &= \frac{C_0}{\pi} \int_0^{t-x/M} \mathrm{d}\tau \int_{y-\nu}^{y+\nu} \frac{\mathrm{d}z}{\sqrt{\nu^2-(y-z)^2}} \\ &= C_0 \int_0^{t-x/M} \mathrm{d}\tau = \frac{C_0}{M}(Mt-x), \\ w &= \int_0^{t-x/M} \psi'(\tau)\,\mathrm{d}\tau = \psi\Big(t-\frac{x}{M}\Big) - C_0, \qquad x < Mt. \end{aligned}$$

Finally, we have

$$d = \frac{1}{b}\Big(C_0 + \psi'\Big(t-\frac{x}{M}\Big)\Big), \qquad u = \frac{1}{b\rho M}\psi'\Big(t-\frac{x}{M}\Big), \qquad v = 0.$$

On the boundary $x = 0$ we obtain

$$\rho u + b\rho v = \frac{C_0}{b} = \frac{1}{b}\int_{\mathbb{R}} y\varphi(0,y)\,\mathrm{d}y = \mathrm{const} \neq 0. \qquad (3.2.21\gamma)$$

Lemma 3.2.3. *Vector* $(3.2.20'\gamma)$ *satisfies system* (3.2.1), *zero initial condition and boundary condition* $(3.2.21\gamma)$.

Clearly, here the consistency condition (3.2.4) is violated.

The solvability set will be determined by the relations

$$N_\gamma \equiv \Big\{\varphi(t,y) \Big/ \int_{\mathbb{R}} \varphi(t,y)\,\mathrm{d}y = 0 \ \cap\ \int_{\mathbb{R}} y\varphi(0,y)\,\mathrm{d}y \neq 0\Big\}.$$

These relations, for instance, are satisfied by the set of functions of the form

$$\varphi(t,y) = T(t)Y(y)\mathrm{e}^{-y^2},$$

where $T(t)$, $Y(y)$ are an arbitrary and an odd functions, respectively.

β) Let us write the problem solution in the semicircle $\{a^2 + b^2 < M^2, a < 0\}$:

$$w(x) = C_\beta(\omega)\hat{q}_j(p, x, \omega).$$

Transform the coefficient

$$C_\beta(\omega) = \frac{p\hat{\varphi}(p,\omega)}{p - i\omega b + aM/r}\bigg|_{p=a\omega} = \frac{M\sigma p\hat{\varphi}(p,\omega)(p - i\omega b - ar/M)}{(p-p_1)(p-p_2)}\bigg|_{p=p_1}.$$

Let us use the representation

$$\hat{\varphi}(p,\omega) = (p - a\omega)\hat{\psi}(p,\omega)/p,$$

where $\hat{\psi}(a\omega;\omega) \neq 0$. The last one in the original takes the form

$$\varphi(t,y) = \int_0^t (\partial_\tau - \beta\partial_y + i\gamma\partial_y)\psi(\tau,y)\,\mathrm{d}\tau.$$

Hence, since $\varphi(t,y)$ is real it follows that $\partial_y\psi = 0$ and $\varphi = \psi$. Therefore, the solvability condition is fulfilled for the function

$$\varphi(t) \equiv \varphi(t,y), \qquad \text{since} \quad \partial_y\varphi(t,y) = 0. \tag{3.2.15β}$$

Let us return to the sought coefficient

$$C_\beta(\omega) = \frac{M\sigma(p - i\omega b - ar/M)\hat{\psi}}{p - p_2}\bigg|_{p=\alpha\omega} = M\sigma\Big[1+\frac{ib}{2\gamma}\Big(\frac{a^2\sigma}{M}+M\sigma-1\Big)\Big]\hat{\varphi}(\alpha\omega),$$

where

$$\hat{\varphi}(\alpha\omega) \underset{\Lambda_t^+}{\overset{\bar{\Delta}}{=}} \int_0^\infty \varphi(t)\mathrm{e}^{-\alpha t\omega}\,\mathrm{d}t \overset{\Phi_\omega^-}{\Rightarrow} \int_0^\infty \frac{\gamma}{\pi}\frac{\varphi(t)t\,\mathrm{d}t}{\gamma^2t^2 + (\beta t + y)^2} = \psi(y),$$

because

$$\Phi_\omega^-(\mathrm{e}^{-\alpha t\omega}) = \int_{\mathbb{R}} \mathrm{e}^{-i\omega(\beta t + y)}\,\mathrm{d}\omega = \frac{1}{\pi}\frac{\gamma t}{\gamma^2t^2 + (\beta t + y)^2}.$$

Thus we obtain

$$u_j(t,x,y) = M\sigma\iint_K \psi(y-z)q_j(t-\tau,x,y-z)\,\mathrm{d}z\,\mathrm{d}\tau, \qquad u_j(0,x,y) = 0,$$

or, componentwise,

$$d = -M\partial_x w,$$

$$u = \frac{M}{\rho}\,\partial_x^2\int_0^t w(\xi,x,y)\,\mathrm{d}\xi, \quad v = \frac{M}{\rho}\,\partial_{xy}^2\int_0^t w(\xi,x,y)\,\mathrm{d}\xi, \tag{3.2.20β}$$

where

$$w(t,x,y) = \iint_K \psi(y-z) q_0(t-\tau, x, y-z)\,\mathrm{d}z\,\mathrm{d}\tau.$$

Let us calculate the limiting values of the components

$$d\big|_{x=0} = -\frac{M}{\pi} d_x \int_0^{1-x/M} \mathrm{d}\tau \int_{y-v}^{y+v} \frac{\psi(y-z)\,\mathrm{d}z}{\sqrt{v^2-(y-z)^2}}\bigg|_{x=0}$$

$$= -\frac{M}{\pi}\,\partial_x \int_0^{1-x/M} \mathrm{d}\tau \int_{-\pi/2}^{\pi/2} \psi(v\sin\varphi)\,\mathrm{d}\varphi\bigg|_{x=0} = -M\psi(0).$$

Finding the integral

$$\int_0^t w(\xi,x,y)\,\mathrm{d}\xi = \frac{1}{\pi}\iint_K \psi(-z)\ln\left|\frac{T+\sqrt{T^2+R^2}}{R}\right|\mathrm{d}z\,\mathrm{d}\tau,$$

where $R^2 = x^2+(y-z)^2$, $T = M(t-\tau)$, and taking into account $\partial_y = -\partial_z$, we obtain

$$v = -\frac{M}{\pi}\,\partial_x \int_0^{t-x/M} \mathrm{d}\tau \left[\psi(y-z)\ln\left|\frac{T+\sqrt{T^2+R^2}}{R}\right|\right]\bigg|_{y-z=\pm v} = 0.$$

By applying the relation $\Box q_0 = 0$ we arrive at

$$u = \frac{M}{\pi\rho}\int_0^t \mathrm{d}\xi \iint_K \psi(y-z)\partial_x^2 q_0\,\mathrm{d}z\,\mathrm{d}\tau$$

$$= \frac{1}{\pi M\rho}\iint_K \psi(y-z)\partial_t q_0(t-\tau,x,y-z)\,\mathrm{d}z\,\mathrm{d}\tau$$

$$- \frac{M}{\pi\rho}\int_0^t \mathrm{d}\xi \iint_K \partial_y^2[\psi(y-z)q_0(t-\tau,x,y-z)]\,\mathrm{d}z\,\mathrm{d}\tau.$$

Determining the first term analogously to d, and the second analogously to v, we obtain

$$u\big|_{x=0} = \frac{1}{\pi M\rho}\,\partial_t \int_0^{t-x/M} \mathrm{d}\tau \int_{-\pi/2}^{\pi/2} \psi(v\sin\varphi)\,\mathrm{d}\varphi\bigg|_{x=0}$$

$$= \frac{1}{\pi M\rho}\int_{-\pi/2}^{\pi/2} \psi(Mt\sin\varphi)\,\mathrm{d}\varphi.$$

Thus, on the boundary ($x = 0$) we have

$$d + a\rho u + b\rho v = -M\psi(0) + \frac{a}{\pi M}\int_{-\pi/2}^{\pi/2} \psi(Mt\sin\varphi)\,\mathrm{d}\varphi = A\int_0^\infty \varphi(t)\frac{\mathrm{d}t}{t}, \tag{3.2.21β}$$

where

$$A = -\frac{M\gamma}{\pi(\gamma^2+\beta^2)} + \frac{a\gamma}{\pi M}\int_{-\pi/2}^{\pi/2} \frac{d\varphi}{\gamma^2+\beta^2+2M\beta\sin\varphi + M^2\sin^2\varphi} \neq 0.$$

The solvability set is determined in the form

$$N_\beta = \Big\{\varphi(t) \Big/ \int_0^\infty \varphi(t)\,\frac{dt}{t} < \infty\Big\}.$$

3.3. LINEARIZED PLANE PROBLEM OF GAS DYNAMICS

Let the supersonic and subsonic stationary gas flows be separated by the surface of a strong discontinuity, i. e., a shock wave (with the equation $x = 0$), and let an exponential perturbation be imposed on the shock wave front at the initial time, whereas the incident flow remains unchanged. The question is: how is the perturbation changing with time? This problem is discussed, in particular, by Blokhin (1982) and Dyakov (1954), where the conditions are found, under which the problem has an unstable solution, i. e., the above-mentioned perturbation becomes very significant and ultimately the shock wave is no longer existent as a physical structure.

The above-described process is mathematically understood as an initial boundary value problem in the half-plane $x > 0$ for the gasdynamic equation with the boundary conditions on the shock wave (Dyakov, 1954):

$$AU_t + BU_x + CU_y = 0, \qquad t > 0, \quad x > 0, \quad y \in \mathbb{R}, \tag{3.3.1}$$

$$U(0,x,y) = 0, \qquad x > 0, \quad y \in \mathbb{R}, \tag{3.3.2}$$

$$LU(t,0,y) = \varphi(t,y), \qquad t > 0, \quad y \in \mathbb{R}. \tag{3.3.3}$$

Here, $A = \operatorname{diag}(m^2, m^2, 1, 1)$, $U = (\delta u, \delta v, \delta p, \delta s)$ are small perturbations of the velocity vector (u, v), pressure p, and entropy s;

$$B = \begin{bmatrix} m^2 & 0 & 1 & 0 \\ 0 & m^2 & 0 & 0 \\ 1 & 0 & 1 & 0 \\ 0 & 0 & 0 & 1 \end{bmatrix}, \qquad C = \begin{bmatrix} 0 & 0 & 0 & 0 \\ 0 & 0 & 1 & 0 \\ 0 & 1 & 0 & 0 \\ 0 & 0 & 0 & 0 \end{bmatrix},$$

$$L = \begin{bmatrix} 1 & 0 & d & 0 \\ 0 & \partial_t & -\lambda\partial_y & 0 \\ 0 & 0 & 1 & 1 \end{bmatrix}, \qquad \varphi = \begin{pmatrix} \varphi_1 \\ \varphi_2 \\ \varphi_3 \end{pmatrix},$$

where $m^2 < 1$, d, λ are physical constants.

Blokhin (1982) analyzed the family of solutions of the form

$$U^0(t,x,y) = U^0 \exp\left[n(\gamma t + ax + iky)\right], \quad i = \sqrt{-1}, \quad n \in \mathbb{N}, \tag{3.3.4}$$

where

$$U^0 \neq 0 \text{ is constant vector}, \qquad \operatorname{Re}\gamma > 0, \quad \operatorname{Re}\alpha < 0, \quad \operatorname{Im} k = 0. \tag{3.3.5}$$

The plane (d, λ) of coefficients of the boundary condition (3.3.3) is subdivided into three parts:

I. The domain of ill-posedness determined by constructing an Hadamard example.

II. The domain, for which exponential solutions damp with time.

III. The domain, for which exponential solutions neither decrease, nor increase.

However, there are calculation errors in work by Blokhin (1982). The main of which is the wrong dispersion relation, i. e., basic equation providing further determination of the solution parameters according to conditions (3.3.5). Therefore, this problem had to be treated again by Temirbulatov and Tepikin (1996), and it is shown that the problem has no Hadamard example of the form (3.3.4) and a new more efficient method of Hadamard example construction is presented.

Further we subdivide the plane (d, λ) into two parts, i. e., the domains of well- and ill-posedness.

3.3.1. Analysis of the boundary condition

Reducing system (3.3.1) to the canonical form (retaining its t-hyperbolicity) we rewrite problem (3.3.1)–(3.3.3) in the following form (Godunov, 1971)

$$U_t + \Lambda U_x + \tilde{C} U_y = 0, \qquad U(0,x,y) = 0, \quad \tilde{\mathcal{L}} U(t,0,y) = \varphi(t,u), \tag{3.3.6}$$

where

$$\Lambda = \operatorname{diag}\left(1, \frac{1+m}{m}, 1, -\frac{1-m}{m}\right), \qquad \tilde{C} = \frac{1}{\sqrt{2}m} \begin{bmatrix} 0 & 1 & 0 & -1 \\ 1 & 0 & 0 & 0 \\ 0 & 0 & 0 & 0 \\ -1 & 0 & 0 & 0 \end{bmatrix},$$

$$\tilde{\mathcal{L}} = \frac{1}{\sqrt{2}m}\begin{bmatrix} 0 & 1+dm & 0 & 1-dm \\ \sqrt{2}\partial_t & -m\lambda d_y & 0 & m\lambda\partial_y \\ 0 & m & \sqrt{2}m & -m \end{bmatrix}.$$

We find from the third equation of the system that

$$U_3(t,x,y) = f(t-x,y). \tag{3.3.7}$$

The arbitrary continuously differentiable function f is determined from the boundary condition

$$f(t,y) = \varphi_3(t,y) + (U_4(t,0,y) - U_2(t,0,y))/\sqrt{2}.$$

By applying integral transforms we obtain the problem for the system of ODE

$$\dot{V} = AV, \quad x > 0, \qquad MV(0) = \tilde{\psi}, \tag{3.3.8}$$

where the vectors $V = (V_1, V_2, V_3)(p,x,\omega)$, $\tilde{\psi} = m(\tilde{\varphi}_1, \tilde{\varphi}_2)(p,\omega)$ and matrices

$$A = \begin{bmatrix} -p & -\dfrac{i\omega}{m\sqrt{2}} & \dfrac{i\omega}{m\sqrt{2}} \\ -\dfrac{i\omega}{\sqrt{2}(1+m)} & -\dfrac{pm}{1+m} & 0 \\ -\dfrac{i\omega}{\sqrt{2}(1-m)} & 0 & \dfrac{pm}{1-m} \end{bmatrix}, \quad M = \begin{bmatrix} 0 & 1+dm & 1-dm \\ p & -\dfrac{i\omega m\lambda}{\sqrt{2}} & \dfrac{i\omega m\lambda}{\sqrt{2}} \end{bmatrix}.$$

The equation

$$\det\|A - vI\| = (p+v)[(1-m^2)v^2 - 2m^2pv - p^2m^2 - \omega^2] = 0$$

has the following roots: let $p \neq \omega$, then

$$v_1 = -p, \qquad v_{2,3} = p(m^2 \mp D)/(1-m^2),$$
$$D = \sqrt{m^2 + (1-m^2)\omega^2/p^2}, \quad \operatorname{Re} D > m > m^2 > 0,$$

hence, if $p = \omega > 0$ ($\omega > 0$, $D = 1$), have

$$v_{1,2} = -\omega, \qquad v_2 = \omega\frac{1+m^2}{1-m^2}.$$

To the numbers $v_1 = -p$, $v_2 = -p(D-m^2)/(1-m^2)$ there correspond the eigenvectors

$$X_1 = \Big(-\frac{i\sqrt{2}p}{\omega}, 1, 1\Big), \quad X_2 = \Big(1, -\frac{i\omega(1-m)}{p\sqrt{2}(m-D)}, \frac{i\omega(1+m)}{p\sqrt{2}(m+D)}\Big), \quad p \neq \omega.$$

To the double root $v = -\omega$ there correspond the eigenvector X_1 and the adjoined vector

$$X_2^0 = \Big(1, \frac{i\omega}{\sqrt{2}p} + \frac{1+m}{p}, \frac{i\omega}{\sqrt{2}p} + \frac{1-m}{p}\Big)\Big|_{p=\omega} = \Big(1, \frac{i}{\sqrt{2}} + \frac{1+m}{\omega}, \frac{i}{\sqrt{2}} + \frac{1-m}{\omega}\Big).$$

A. Substituting the general bounded solution of differential equation of (3.3.8)

$$V = \tilde{C}_1 X_1 \mathrm{e}^{-px} + \tilde{C}_2 X_2 \mathrm{e}^{v_2 x}, \qquad \operatorname{Re} v_2 < 0,$$

into the boundary condition we arrive at the system

$$\tilde{C}_1 + \frac{i\omega}{p\sqrt{2}(D^2 - m^2)}\big[D - m^2 + dm^2(1-D)\big]\tilde{C}_2 = \frac{\tilde{\psi}_1(p,\omega)}{2},$$
$$-\frac{i\sqrt{2}p^2}{\omega}\tilde{C}_1 + \Big[p + \frac{\lambda m^2\omega^2}{p(D^2 - m^2)}(1-D)\Big]\tilde{C}_2 = \tilde{\psi}_2(p,\omega).$$

Equating to zero the determinant of the latter, i. e., assuming

$$\frac{m^2\omega^2(1-D)}{p(D^2-m^2)}\Big(\lambda - \frac{p^2 d}{\omega^2} - \frac{p^2 D}{m\omega^2}\Big) = 0,$$

we obtain the irrational equation

$$z\sqrt{1 + \frac{2\lambda_0}{z}} = \lambda - dmz, \qquad \lambda_0 = \frac{1-m^2}{2m^3} > 0,$$

where $z = p^2/(m\omega^2) > 0$. Squaring both sides we arrive at the square equation

$$az^2 + 2bz - \lambda^2 = 0,$$

where $a = 1 - d^2m^2$, $b = \lambda_0 + dm\lambda$.

First we will calculate the positive roots of the latter and then find out, which of those is the roots of the irrational equation. By Vieta formulas

$$z_1 z_2 = -\lambda^2/a, \qquad z_1 + z_2 = -b/(2a).$$

In the general case, i. e., when $ab\lambda \neq 0$, we yield

1. $z_1^+ = (\Delta - b)/a > 0$, $\Delta = b^2 + a\lambda^2$, if $a > 0$.

2. $z_{1,2} = (\pm\Delta - b)/a > 0$, if $a < 0$ and $b > 0$.

Let us consider particular cases:

3. $a = 0$, $b\lambda \neq 0$, there are two roots

$$z^{\pm} = \begin{cases} \lambda^2/(2(\lambda_0 + \lambda)) > 0, & \lambda > -\lambda_0, \ dm = 1, \\ \lambda^2/(2(\lambda_0 - \lambda)) > 0, & \lambda < \lambda_0, \ dm = -1. \end{cases}$$

4. $b = 0$, $a\lambda \neq 0$, there is single root

$$z_1 = \lambda^2/\sqrt{\lambda^2 - \lambda_0^2}, \qquad \lambda^2 > \lambda_0^2, \quad dm = -\lambda_0/\lambda.$$

5. $\lambda = 0$, $ab \neq 0$, there are the following roots

$$z_1^0 = 2\lambda_0/(d^2m^2 - 1) > 0, \qquad z_2 = 0.$$

Now, let us find out whether squaring resulted in an extraneous root. Verification shows that z_1^0 (the case $\lambda = 0$) will not be the root of the basic equation.

Further, we study the locus on the plane (λ, d) of the relations connected with the remaining roots. Thus:

1. $\Rightarrow$ the open band $-1/m < d < 1/m$ with the cut along the interval $(-1/m, 1/m)$ along the axis $\lambda = 0$.

2. $\Rightarrow$ halfplanes $(-\lambda_0 < \lambda \cap d > 1/m)$ and $(\lambda < \lambda_0 \cap d < -1/m)$.

3. $\Rightarrow$ halflines $(d = 1/m,\ \lambda > -\lambda_0)$ and $(d = -1/m,\ \lambda < \lambda_0)$.

4. $\Rightarrow$ domain, determined by the hyperbola $d\lambda = -\lambda_0/m$, where $\lambda > \lambda_0$ and $\lambda < -\lambda_0$ belongs to the domains from 1 and 2.

Lemma 3.3.1. *Problem* (3.3.1)–(3.3.3) *is well posed in two quarters* $(\lambda_0 \leq \lambda \cap d \leq -1/m)$, $(\lambda \leq -\lambda_0 \cap d \geq 1/m)$ *and interval* $-1/m < d < 1/m$, $\lambda = 0$. *Their complement to the whole plane* (λ, d) *comprises the domain of ill-posedness.*

B. Consider the case $\nu_{1,2} = -p$. Substituting vector (solution)

$$V = \tilde{C}_1 X_1^0 \mathrm{e}^{-px} + \tilde{C}_2 (X_2^0 + xX_1)\mathrm{e}^{-px}, \tag{3.3.9}$$

in the boundary condition (3.3.7) ($x = 0$, $p = \omega > 0$) we obtain

$$\tilde{C}_1 + \Big(\frac{i\omega}{\sqrt{2p}} + \frac{(1+dm^2)p}{\omega^2}\Big)\tilde{C}_2 = \frac{\tilde{\psi}_1(\omega,\omega)}{2},$$
$$-\frac{pi\sqrt{2}\tilde{C}_1}{\omega} + \Big(1 - \frac{ic}{\omega^2}\Big)\tilde{C}_2 = \frac{\tilde{\psi}_2(\omega,\omega)}{p},$$
$$c = \sqrt{2}\lambda m^2. \tag{3.3.10}$$

The matrix of the system is degenerate along the line $\lambda - d - 1/m^2 = 0$, lying in the previous domain of ill-posedness.

The above analysis shows that the condition of ill-posedness is of "coefficient and root" type and it can be written in the form

$$\mathrm{I} \equiv (p = a|\omega|, \quad a = \sqrt{mz_j} > 0).$$

3.3.2. Solution of the ill-posed problem (particular case)

Let us investigate problem (3.3.1)–(3.3.3) in the domain of ill-posedness (see Lemma 3.3.1). Temirbulatov and Tepikin (1996) proved its conditional well-posedness out of the straight line $1 + dm = 0$ by reducing it to the operator Volterra equation of the second kind using well-known method of characteristics and defined a pseudosolution along this line. The case $p = \omega$, for which the I-condition is of the form

$$\mathrm{I}' \equiv [(p = w) \cap (\lambda - d - 1/m^2 = 0)]$$

is not considered in the paper.

a) Solution of SLAE (3.3.10). Under I$'$-condition system (3.3.10) can be reduced to the equation

$$\sqrt{2}\tilde{C}_1 + (i + c/\omega)\tilde{C}_2 = \tilde{\psi}_1(\omega,\omega)/\sqrt{2} \tag{3.3.10$'$}$$

with the additional requirement

$$\tilde{\psi}_2(\omega,\omega) = -i\omega\tilde{\psi}_1(\omega,\omega). \tag{3.3.11$'$}$$

Find the normal solution

$$\tilde{C}^n(\omega) = (i\omega\sqrt{2}/\Omega, i\Delta/\omega)\tilde{\psi}_2(0,\omega),$$

where $\Delta = c + i\omega$, $\Omega = \Delta^2 + 2\omega^2$, $c = \sqrt{2}\lambda m^2 \neq 0$. When $c = 0$ ($\lambda = 0$), we have

$$\tilde{C}^n(\omega) = (i\sqrt{2}/\omega, -1/\omega)\tilde{\psi}_2(\omega,\omega).$$

Obviously, $\tilde{C}^n$ is the exact solution of (3.3.10) with the solvability condition (3.3.11′).

b) Solution of the problem for ODE. As is easily seen, the vector V^n (3.3.9), where $\tilde{C}^n$ acts as a coefficient, is also the exact solution of problem (3.3.8), but only provided the solvability condition (3.3.11′) and ill-posedness I-condition are fulfilled.

c) Solution of the original problem. We have

$$\tilde{\varphi}(\omega,\omega) = \tilde{\varphi}(p,\omega)\big|_{p=\omega} \overset{\Phi_{\bar{\omega}}(y)}{\Rightarrow} \bar{\varphi}(\omega,y) = \int_0^\infty \varphi(t,y)\mathrm{e}^{-\omega t}\,\mathrm{d}t.$$

Therefore, from (3.3.11′) we get the equality

$$\varphi_2(t,y) = -\partial_y\varphi_1(t,y), \quad \text{or} \quad \varphi_1(t,y) = -\int_0^y \varphi_2(t,z)\,\mathrm{d}z, \tag{3.3.11}$$

expressed via the vector of the right-hand side of boundary condition (3.3.3).

Continuing the integral inversion we yield

$$\int_0^\infty \varphi(t,y)\mathrm{e}^{-\omega t}\,\mathrm{d}t \overset{\Phi_{\bar{\omega}}}{\Rightarrow} \int_0^\infty \varphi(t,y)\mathrm{e}^{-\omega t}\,\mathrm{d}t\Big(\frac{1}{\pi}\int_0^\infty \mathrm{e}^{-\omega(t+iy)}\,\mathrm{d}\omega\Big)$$
$$= \frac{1}{\pi}\int_0^\infty \frac{\varphi_k(t,y)\,\mathrm{d}t}{t+iy} \equiv a_k(y) - iyb_k(y). \tag{3.3.12}$$

For the function to remain real, the following equalities must be satisfied

$$a_k = \frac{m}{\pi}\int_0^\infty \frac{\varphi_k(t,y)}{t^2+y^2}\,t\,\mathrm{d}t, \quad \int_0^\infty \frac{\varphi_k(t,y)}{t^2+y^2}\,\mathrm{d}t = 0, \quad b_k = 0, \quad k = 1,2. \tag{3.3.13}$$

Originals of functions. Let us rewrite the vector V^n in the form more convenient for the calculation

$$v_1^n = \mathrm{e}^{-px}\Big(\frac{2p}{\Omega} + x\frac{\sqrt{2}\,\Delta}{\omega\Omega}p + \frac{i\Delta}{\Omega}\Big)\tilde{\psi}_2(\omega,\omega),$$
$$v_{2,4}^n = \mathrm{e}^{-px}\Big(\frac{i\omega\sqrt{2}}{\Omega} + x\frac{i\Delta}{\Omega} + \Big(\frac{i\omega}{\sqrt{2}} + 1 \pm m\Big)\frac{i\Delta}{\Omega}\Big)\tilde{\psi}_2(\omega,\omega), \quad c \neq 0. \tag{3.3.9.1}$$

If $c = 0$ then

$$v_1^0 = \mathrm{e}^{-px}\tilde{\psi}_2\Big(\frac{2+i\sqrt{2}x}{\omega^2}, p - \frac{1}{\omega}\Big) = p\mathrm{e}^{-px}\tilde{\psi}_1\Big(\frac{\sqrt{2}x}{\omega} + i\Big(1 - \frac{2}{\omega}\Big)\Big),$$
$$v_{2,4}^n = \mathrm{e}^{-px}\tilde{\psi}_2\Big(\frac{i\sqrt{2}-x}{\omega} - \frac{1}{p}\Big(\frac{i}{\sqrt{2}} + 1 \pm m\Big)\frac{i\Delta}{\Omega}\Big). \tag{3.3.9.2}$$

By the theorem on the convolution of two functions we have

$$u_j^n(t,x,y) = \int_0^t \mathrm{d}\tau \int_0^\infty \varphi(z) G_j(t-\tau, x, y-z)\,\mathrm{d}z, \tag{3.3.14}$$

where $\varphi(z)$ is determined by (3.3.12) and (3.3.13), and the kernels G_j are to be found.

Let us perform auxiliary calculations

1. The inverse Laplace transform for $x \geq 0$:

$$\mathrm{e}^{-px} \Rightarrow \delta(t-x),$$
$$p\mathrm{e}^{-px} = -\partial_x \mathrm{e}^{-px} \Rightarrow \partial_x\delta(t-x) = \partial_t\delta(t-x) = \delta', \mathrm{e}^{-px}/p \Rightarrow \theta(t-x),$$

$\delta(t)$ is the Dirac delta-function ($\delta(t) = 0$, $t \neq 0$; $\delta(0) = \infty$), $\theta(t)$ is the Heaviside function ($\theta(t) = 1$, $t > 0$; $\theta(t) = 0$, $t \leq 0$).

2. The expansion in simple fractions

$$\frac{1}{\Omega} = \frac{1}{(\alpha+i\omega)(\beta-i\omega)} = \frac{1}{2\sqrt{2}\,c}\Big(\frac{1}{\alpha+i\omega} + \frac{1}{\beta+i\omega}\Big),$$
$$\frac{\Delta}{\Omega} = \frac{1}{2}\Big(\frac{1}{\beta-i\omega} - \frac{1}{\alpha+i\omega}\Big),$$

where $\alpha = (\sqrt{2}+1)c$, $\beta = (\sqrt{2}-1)c$.

3. The Fourier inverse transform

$$\frac{1}{a \pm i\omega} \Rightarrow \mp 2\,\mathrm{sign}\,(y)\mathrm{e}^{-|ay|}, \quad ay \neq 0, \quad \mathrm{sign}\,y = \begin{cases} 1, & y > 0, \\ -1, & y < 0, \end{cases}$$
$$\frac{1}{\omega} \Rightarrow \frac{2}{\pi}\int_0^\infty \frac{\sin(y\omega)}{\omega}\,\mathrm{d}\omega = 1.$$

This implies

$$\frac{1}{\Omega} \Rightarrow \frac{1}{\sqrt{2}c}\varphi^-(y), \qquad \frac{\Delta}{\Omega} \Rightarrow \varphi^+(y), \qquad \frac{\Delta}{\omega\Omega} \Rightarrow \varphi_0^+,$$

where

$$\varphi^{\pm}(y) = \operatorname{sign}(y)(\mathrm{e}^{-|\beta||y|} \pm \mathrm{e}^{-|\alpha||y|}), \qquad \varphi_0^+ = \int_0^\infty \varphi^+(y)\,\mathrm{d}y.$$

4. The construction of the kernels G on the basis of formulas (3.3.9.1) and (3.3.9.0)

$$\begin{aligned} G_1 &= \sqrt{2}(\varphi^-(y)/c + x\varphi_0^+)\delta'(t-x) + i\varphi^+(y)\delta(t-x) \\ &= \sqrt{2}\varphi^-(y)\delta'/c + \sqrt{2}\varphi_0^+\delta + i\varphi^+(y)\delta, \\ G_{2,4} &= (\sqrt{2}/c)\partial_y\varphi^-(y) + i[x\delta + \theta(1 \pm m + (1/\sqrt{2})\partial_y)]\varphi^+(y), \\ G_1^0 &= \sqrt{2}x\delta' + i[\delta - 2\delta'] = \sqrt{2}\delta + i(\delta - 2\delta'), \\ G_{2,4}^0 &= -[x\delta + \theta(1 \pm m)] + i(2\delta - \theta)/\sqrt{2}. \end{aligned}$$

5. The components u_j^n is calculated using formula (3.3.14) proceeding $\operatorname{Re} G_j$. The resulting equalities are valid in the domain $0 < x < t$, $y \in \mathbb{R}$, for $x \geq t$ the components are zero. The latter is consistent with the influence of the zero initial condition

$$\begin{gathered} u_1^n(t,x,y) = \sqrt{2}a_2^0\varphi_0^+ \equiv \text{const}, \qquad u_1^0(t,x,y) = \sqrt{2}a_1^0 \equiv \text{const}, \\ u_{2,4}^n(t,x,y) = \partial_y\varphi_2^{0-}(y) \equiv u_{2,4}^n(y), \\ u_{2,4}^0(t,x,y) = -a_2^0[t \pm m(t-x)] \equiv u_{2,4}^0(t,x), \end{gathered}$$

where

$$a_k^0 = \int_0^\infty a_k(z)\,\mathrm{d}z, \quad k = 1,2, \qquad \varphi_2^{0-}(y) = \frac{1}{c}\int_0^\infty a_2(z)\varphi^-(y-z)\,\mathrm{d}z.$$

Let us determine more precisely the component u_3^n:

$$\begin{gathered} u_3^n(t,x,y) = \varphi_3(t-x,y)/\sqrt{2}, \\ u_3^0(t,x,y) = \varphi_3(t-x,y)/\sqrt{2} + \sqrt{2}m(t-x)a_2^0, \qquad 0 < t < x. \end{gathered}$$

6. Finally, let us determine which problem is solved by the vector u^n (or, to be more precise, which equalities are true).

By the support of the components there immediately follows the fulfillment of the homogeneous initial condition (3.3.2)

$$u^n(t,x,y) = 0, \qquad t = 0, \quad x > 0, \quad y \in \mathbb{R}.$$

Table 3.1

u_j	$x=0$	$\partial_t+\partial_x$	∂_x	∂_y
u_1	$\sqrt{2}a_2^0\varphi_0^+$	0	—	0
u_1^0	$\sqrt{2}a_1^0$	0	—	0
$u_{2,4}(y)$	$\partial_y\varphi_2^{0-}(y)$	0	0	$\partial_y^2\varphi_2^{0-}(y)$
$u_{2,4}^0(t,x)$	$-t(1\pm m)a_2^0$	$-a_2^0$	$\pm ma_2^0$	0
u_3	$\varphi_3(t,y)/\sqrt{2}$	0	—	—
u_3^0	$\varphi_3(t,y)/\sqrt{2}+\sqrt{2}mta_2^0$	0	—	—
u_2+u_4	$2\partial_y\varphi_2^{0-}(y)$	—	—	—
$u_2^0+u_4^0$	$-2a_2^0t$	—	—	—
u_2-u_4	0	—	—	0
$u_2^0-u_4^0$	$-2ma_2^0t$	—	—	0

For clarity, we gather all results in Table 3.1, from which follows

Statement 3.3.1. If

$$\int_0^\infty \frac{\varphi_k(t,y)\,\mathrm{d}y}{t^2+y^2}=0, \qquad k=1,2,$$

where $\varphi_2(t,y)=-\partial_y\varphi_1(t,y)$, then the vector $u^n(t,x,y)$ on the straight line $\lambda-d=1/m^2$ satisfy the system, initial condition, and the third equality of the boundary condition (3.3.6), and the remaining two equalities have other right-hand sides.

Bibliography

Basarbayeva S. E. (1982). *On the solvability of mixed problems for the system of differential equations of heat and mass exchange.* Candidate Degree work. Almaty (in Russian).

Bateman H. and Erdélyi A. (1969). *Tables of Integral Transforms. Vol. 1.* Nauka, Moscow (in Russian).

Bellman R. E. (1970). *Introduction to Matrix Analysis.* McGraw-Hill, New York.

Bitsadze A. V. (1976). *Equations of Mathematical Physics.* Nauka, Moscow (in Russian).

Blokhin A. M. (1982). *Energy Integrals in the Problem of the Shock Wave Stability.* Nauka, Novosibirsk (in Russian).

Bukhgeim A. L. (1999). *Volterra Equations and Inverse Problems.* VSP, Utrecth.

Ditkin V. A. and Prudnikov A. P. (1974). *Integral Transforms and Operator Calculus.* Nauka, Moscow (in Russian).

Dyakov S. P. (1954). On shock waves stability. *Zh. Expir. Teor. Fiziki* 27 (3), 288–295 (in Russian).

Eidelmann S. D. (1964). *Parabolic systems.* Nauka, Moscow (in Russian).

Fedosov V. P. and Yanenko N. N. (1982). Solution of a mixed problem for the wave equation. *Dokl. Akad. Nauk SSSR* 262 (3), 549–553 (in Russian).

Friedman A. (1968). *Partial differential] equations of a parabolic type.* Mir, Moscow (in Russian).

Gantmacher F. R. (1966). *Theory of Matrices.* Nauka, Moscow (in Russian).

Godunov S. K. (1971). *Equations of Mathematical Physics.* Nauka, Moscow (in Russian).

Godunov S. K. (1983). *Matrix Exponent, Green's Matrix, and Lopatinckii's Condition.* Novosibirsk State University, Novosibirsk (in Russian).

Godunov S. K. (1986). *Collected Problems on Ordinary Differential Equations.* Novosibirsk State University, Novosibirsk (in Russian).

Godunov S. K. (1994). *ODE with Constant Coefficients. Vol. 1. Boundary-Value problems.* Novosibirsk State University, Novosibirsk (in Russian).

Godunov S. K. and Gordiyenko V. M. (1977). Mixed problem for the wave equation. *Proceedings of S. L. Sobolev's Seminar.* Nauka, Novosibirsk, 5–31 (in Russian).

Gordiyenko V. M. (1981). Symmetrization of a mixed problem for a second-order hyperbolic equations with two space variables. *Sib. Mat. Zh.* 22 (2), 84–104 (in Russian).

Gradstein I. S. and Ryzhik I. M. (1971). *Tables of Integrals, Sums, Series, and Products.* Nauka, Moscow (in Russian).

Hadamard J. (1902). Sur les problems aux derivees partielles et leur signification physique. *Bull. Univ. Princeton* 13.

Hadamard J. (1952). *Lectures on Cauchy's Problem in Linear Partial Differential Equations.* Dover, New York.

Ivanov V. K. (1962). On linear ill-posed problems. *Dokl. Akad. Nauk SSSR* 145 (2), 270–272 (in Russian).

Ivanov V. K. (1963). On ill-posed problems. *Mat. Sbornik* 61 (2), 211-223 (in Russian).

John F. (1955). A note on "improper" problems in partial differential equation. *Comm. Pure Appl. Math.* 8 (4), 591–594 (in Russian).

Kasymov D. K. (1999). *A method for the solution of a boundary-value problem for a system of linear differential equations. Vestnik KazGU. Ser. Mat., Mekh. i Inf.* 4, 37–43 (in Russian).

Kreiss H. O. (1970). Mixed problem for hyperbolic systems. *Matematika* 14 (4), 98–115 (in Russian).

Lavrent'ev M. M. (1956). Cauchy problem for Laplace equation. *Izv. Akad. Nauk SSSR. Ser. Mat.* 20, 819–842 (in Russian).

Lavrent'ev M. M. (1962). *Some Ill-Posed Problems of Mathematical Physics.* Nauka, Novosibirsk (in Russian).

Ladyzhenskaya O. A., Uraltseva N. N., and Solonnikov V. A. (1967). *Linear and Quasi-Linear Equations of a Parabolic Type.* Nauka, Moscow (in Russian).

Lancaster P. (1978). *Theory of Matrices.* Nauka, Moscow (in Russian).

Liskovets O. (1982). Theory and methods for the solution of ill-posed problems. *Itogi Nauki i Tekh. Ser. Mat. Anal.* 20, 116–178 (in Russian).

Lopatinskii Ya. B. (1984). *Theory of General Boundary-Value Problems.* Nauka, Kiev (in Russian).

Mikhailov V. P. (1959). On potentials of parabolic equations. *Dokl. Akad. Nauk SSSR* 129 (3), 1226–1229 (in Russian).

Mikhailov V. P. (1960). Solution of a mixed problem for a parabolic system by the method of potentials. *Dokl. Akad. Nauk SSSR* 132 (2), 291–294 (in Russian).

Morozov V. A. (1987). *Regular Methods for the Solution of Ill-Posed Problems.* Nauka, Moscow (in Russian).

Novikov P. S. (1938). On the solution uniqueness of the inverse problem of the potential theory. *Dokl. Akad. Nauk SSSR* 18, 165–168 (in Russian).

Petrovskii I. G. (1938). On Cauchy problem for system of linear partial differential equations in the field of nonanalytic functions. *Bull. Mos. Gos. Univ. Sec. A.* 1 (7), 1–72 (in Russian).

Pontryagin L. S. (1988). *Differential Equations and their Applications.* Nauka, Moscow (in Russian).

Sakamoto R. (1972). Mixed problems for hyperbolic equations I. Energy inequalities. In: *Matematika* 16 (1), 62–100 (in Russian).

Smirnov V. I. (1953). *A Course of Higher Mathematics. Vol. 3, Pt 1.* Nauka, Moscow (in Russian).

Smirnov V. I. (1957). *A Course of Higher Mathematics. Vol. 3, Pt 2.* Nauka, Moscow (in Russian).

Solonnikov V. A. (1965). Boundary-value problems for linear parabolic systems of general differential equations. *Trudy MIAN Steklova* 83 (in Russian).

Temirbolat S. E. (1998). Algebraic aspects of ill-posed boundary-value problems. *Proceedings of conf. on mathematics dedicated to the memory of academician A. D. Taymanov.* Almaty, 48–54 (in Russian).

Temirbolat S. E. (1998). Are there many ill-posed boundary-value problems? *Abstracts of intern. conf. Inverse Problems of Mathematical Physics.* Novosibirsk (in Russian).

Temirbolat S. E. (1999). On the theory of parabolic problems. *Abstracts of intern. conf. Differential Equations, Analysis of Algebra Problems.* Aktoebe (in Russian).

Temirbolat S. E. (2001). On the first boundary-value problem for parabolic systems (S. D. Eidelmann's hypothesis, V. A. Solonnikov's examples). *Dokl. Nac. Akad. Nauk Resp. Kaz.* 3, 5–14 (in Russian).

Temirbolat S. E. (2001). New method for the solution of ill-posed boundary-value problems. *Dokl. Nac. Akad. Nauk Resp. Kaz.* 4, 11–15 (in Russian).

Temirbolat S. E. (2001). Solution of conditionally well-posed problem of mass and heat exchange. *Vestnik KazNU. Ser. Mat.* 5, 59–67 (in Russian).

Temirbolat S. E. (2002). New approach to conditionally well-posed parabolic problems. *Proceedings of intern. conf. Differential Equations and their Applications.* Almaty, 133–136 (in Russian).

Temirbolat S. E. and Imangaliyeva Yu. V. (2002). An ill-posed boundary-value problem for the first-order ODE. *Vestnik KazNU. Ser. Mat., Mech. i Inf.* 2, 107–112 (in Russian).

Temirbulatov S. I. (1980). Mixed problem for the wave equation. *Diff. Uravn.* 16 (3), 532–536 (in Russian).

Temirbulatov S. I. (1980). Problem with an oblique derivative for the wave equation. *Sib. Mat. Zh.* 21 (5), 78–87 (in Russian).

Temirbulatov S. I. (1989a). Research in the theory of conditionally well-posed problems for parabolic equations. Degree work. Almaty (in Russian).

Temirbulatov S. I. (1989b). A method for studying conditionally well-posed problems. *Vestnik Akad. Nauk KazSSR* 8, 62–64 (in Russian).

Temirbulatov S. I. (1993). An approach to Lopatinckii's condition. *Dokl. Nac. Akad. Nauk Resp. Kaz.* 4, 3–8 (in Russian).

Temirbulatov S. I. (1995). A conditionally well-posed problem for a system of heat and mass exchange. *Vestnik Nac. Akad. Nauk Resp. Kaz.* 2, 73–77 (in Russian).

Temirbulatov S. I. (1995). Three "conditions" of ill-posed problems analysis for a certain class of differential equations. *Proceedings of school-seminar on Mathematics and Mechanics dedicated to 60 anniversary of corresponding member of National Acad. Sci. of Resp. Kazakhstan K. A. Kasymov.* 138–139 (in Russian).

Temirbulatov S. I. (1996). Triads in the theory of ill-posed problems. *Proceedings of science-methodical seminar dedicated to 50 anniversary of professor E. Y. Bidaybekov.* Almaty State University, Almaty, 45–60 (in Russian).

Temirbulatov S. I. (1996). Construction of ill-posed boundary-value problems. *Abstracts of intern. conf. Actual Problems of Mathematics and Mathematical Modelling of Ecological Systems dedicated to 60 anniversary of academician U. M. Sultangazin.* Almaty, p. 31 (in Russian).

Temirbulatov S. I. (1996). *Solution Methods for Ill-Posed Boundary-Value Problems.* Almaty State University, Almaty (in Russian).

Temirbulatov S. I. and Tepikin P. B. (1996). Linearized plane ill-posed problem of gas dynamics. *Izv. MN Acad. Nauk Resp. Kaz. Ser. Fiz. i Mat.* 3, 44–52 (in Russian).

Temirbulatov S. I. and Turdiyeva G. T. (1998). Ill-posed problems of ODE systems and linear algebraic equations. *Proceedings of conf. on mathematics dedicated to the memory of academician A. D. Taymanov.* Almaty, 55–60 (in Russian).

Tikhonov A. N. (1943). On stability of inverse problems. *Dokl. Akad. Nauk SSSR* 39 (5), 195–198 (in Russian).